T0291130

COSMIC
RAY
MUOGRAPHY

COSMIC RAY MUOGRAPHY

Editors

Paola Scampoli
University of Napoli Federico II, Italy & University of Bern, Switzerland

Akitaka Ariga
Chiba University, Japan & University of Bern, Switzerland

World Scientific

NEW JERSEY · LONDON · SINGAPORE · BEIJING · SHANGHAI · HONG KONG · TAIPEI · CHENNAI · TOKYO

Published by

World Scientific Publishing Co. Pte. Ltd.

5 Toh Tuck Link, Singapore 596224

USA office: 27 Warren Street, Suite 401-402, Hackensack, NJ 07601

UK office: 57 Shelton Street, Covent Garden, London WC2H 9HE

Library of Congress Control Number: 2023933136

British Library Cataloguing-in-Publication Data
A catalogue record for this book is available from the British Library.

COSMIC RAY MUOGRAPHY

ISBN 978-981-126-490-0 (hardcover)
ISBN 978-981-126-491-7 (ebook for institutions)
ISBN 978-981-126-492-4 (ebook for individuals)

For any available supplementary material, please visit
https://www.worldscientific.com/worldscibooks/10.1142/13102#t=suppl

Typeset by Stallion Press
Email: enquiries@stallionpress.com

In memory of Peter Grieder

Contents

Chapter 2. Cosmic Ray Muons 33

Peter K. F. Grieder

Chapter 3. Principle of Cosmic Muography — Techniques and Review 85

Paolo Checchia

Chapter 4. Emulsion Detectors for Muography 119

Akira Nishio

Chapter 7. Muography and Geology: Volcanoes, Natural Caves, and Beyond 189

Jacques Marteau

Chapter 8. Muography and Geology: Alpine Glaciers 215

Ryuichi Nishiyama

Chapter 1

Historical Developments and Perspectives of Muography

Hiroyuki K. M. Tanaka

University of Tokyo, Tokyo, Japan.
ht@eri.u-tokyo.ac.jp.

Muography has been successfully applied to the study of gigantic solid earth targets, in particular focusing on subsurface regions including geology surveys and prediction of natural disasters (such as volcanic eruptions and underground water associated disasters) with resolutions at the meter scale. The most essential key to developing this technique is to make improvements in the speed and accuracy of the imaging of these phenomena, concentrating specifically on subsurface geofluid flow. In this chapter, technological developments of muography: from static to dynamic, from 2D to 3D, from station-based to portable, from land-based to airborne, and from natural intelligence to artificial intelligence, will be discussed. In addition, muography art (art inspired by muography) will be introduced, focusing on how pioneering collaborations between scientists and artists have started to make positive impacts on local communities.

1. Introduction

About 13 years after George (1955)[1] first proposed and implemented densimetric measurements of the rock overburden above a gallery tunnel of the Snowy Mountains Hydro Electric Authority in Australia with cosmic-ray muons, particle detection techniques had been developed to the extent that Alvarez *et al.* (1970)[2] could successfully conduct the first muography experiment by applying spark chambers with digital readout units. The term muography literally means "muon rendering" in ancient Greek. While photography utilizes photos (the word for "light" in ancient Greek), muography utilizes the characteristics of relativistic muons. Muography can be explained by using the analogy of medical radiographic imaging that utilizes relativistic particles as probes. The muon, also indicated by the Greek letter μ, is a charged lepton, which is free from strong force interaction but sensitive to electromagnetic force. This particle has the same properties as an electron, but its mass is approximately 200 times larger than the electron. Therefore, relativistic muons have a stronger penetration power. By taking advantage of this unique property of muons, muography creates shadows similar to radiography inside hectometer- to kilometer-scaled objects. Muons can also be used to render the centimeter-scaled heavy (high-Z) objects by following and tracking the single-scattered muons. However, since its application range is limited to smaller objects in comparison to the regular geophysical target size and, thus, there have not been successful applications of small-sized objects in the field of earth, planetary, environmental sciences, or engineering so far, muographic targets smaller than 10 m will not be discussed here.

In this chapter, the pioneering works and future prospects of muography are introduced in 2 sections focusing on 6 topics and 2 topics, respectively; Section 1: (A) early works in muography, (B) revealing the internal structure of a volcano, (C) real-time monitoring and multiply-exposed muography, (D) developments from 2D projection imaging to 3D reconstruction, (E) mobile muography, (F) usage of AI, and (G) muography art; and Section 2: (A) extension from land-based muography to ocean-based muography, and (B) from the terrestrial muography to outer-space muography. Readers can find more detailed and technical descriptions about the recent developments of detectors, techniques, and applications in the following chapters: Chapters 4 and 5: "Emulsion detector for muography" and "Real time detector for muography"; Chapter 6: "Three-dimensional muography and image reconstruction using the filtered back-projection method", for the tomographic image reconstruction technique; Chapter 7: "Muography and geology: Volcanoes, natural caves, and beyond", for volcano and natural cave applications; Chapter 8: "Muography and geology: Alpine glaciers", for alpine glacier applications; Chapter 9: "Muography and archaeology", and for pyramid applications and Chapter 10 of the present issue: "Civil and industrial applications of muography".

2. Historical Developments and Pioneering Works

2.1. *Early works*

George (1955)[1] compared the muon flux inside and outside of the Australian Guthega–Munyang tunnel to calculate the muon's transmission rate and measured the densimetric thickness of the rock overburden. His muonic results were compared with the drilling core sampling results, and they were in agreement within the error bars. His measurement was conducted with a Geiger counter; thus, he could not image the density structure of the rock overburden, but his experiment was important to later motivate Luis Alvarez to attempt the first muography experiment at the Egyptian pyramid.

After the invention of a spark chamber with a digital readout, imaging became more realistic. Alvarez *et al.* (1970)[2] recorded the trajectories of muons inside the pyramid and studied their transmission through the pyramid to establish why the Pyramid of Chephren had only one burial chamber (the so-called Belzoni chamber) while the older-pyramid made by his father Cheops had more complicated internal structures, including the King's and Queen's chambers and the Grand Gallery. This joint project between the United States and the United Arab Republic began in 1966. They constructed their observation system inside the Belzoni chamber of the Chephren Pyramid, and their entire system weighed more than 10 tons. Despite the fact that no hidden rooms were found inside the pyramid, this group showed the potential of the muography technique by profiling the external shape of the Chephren pyramid from the Belzoni Chamber. This pioneering work paved the way for the application of muography in various fields.

After Alvarez's trial, there was a long silence in research activities until Nagamine *et al.* (1995)[3] and Caffau *et al.* (1997)[4] independently installed their muography detectors near geological targets in Japan (mountain) and Italy (Karst topography) to calibrate their muographic detectors by profiling the topographic shapes of

4 H. K. M. Tanaka

these objects along the muon paths. Both of these groups confirmed that the muon flux after passing through these target objects was accurately reflecting the terrain topography of these target volumes that were much larger than the Chephren pyramid. The first attempt to combine the muographic data with other geophysical data was made by Caffau *et al.* (1997).[4] Since muography gives integrated information only along the muon paths, the structural information along the muon paths has to be solved by adding other directional information. They measured the volume and mass of a red-soil deposit laying beneath the sinkhole by combining the near horizontal information (muographic data) with the vertical information (gravimetric data).

2.2. *Revealing the internal structure of a volcano*

2.2.1. *Phreatic explosion*

Muography offers an alternative method to understand the process of the explosion of the volcanic plug by over-pressurized gas.[5] The first successful muographic observation conducted in 2006 captured the shape of the magma deposit generated in 2004 inside the crater (Fig. 1(a)).[6] The thickness of this volcanic plug was less than 100 m and was in agreement with airborne synthetic aperture radar (SAR) measurements conducted both before and after the 2004 eruption.[7] Additionally, a 2006 measurement of the same volcano imaged a low-density region underneath the crater floor. This low-density region was interpreted as a porous magma pathway that was plugged by magma deposited on the crater floor and was created by the following process: after the 2004 eruption process concluded, the magma deposit on the crater floor cooled and solidified, and the magma in the pathway drained away. If this space is over-pressured by future eruptions, the plug may explode, releasing fragments of this magma deposit.

Muographic monitoring of Asama volcano has been ongoing since 2008 after the installation of a scintillator-based muographic observation system in a vault created 1.2 km east of the center of the crater.[6] During the observation, Asama erupted on Feb 2, 2009.

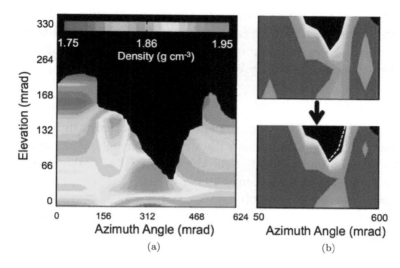

Figure 1. Muographic image of the summit region of Asama volcano, Japan (a) and the time-sequential images before and after the 2009 eruption (b). The dashed line in (b) indicates the shape of the crater before the eruption. The green-yellow-red region in (a) indicates the magma deposit in the crater and the low-density region (a blue patch) was interpreted as a porous magma pathway that was plugged by the magma deposit.

These first time-sequential muographic images (Fig. 1(b)) captured this eruption event and showed that the average density of the northern section of the magma deposit along the muon path was lower after the eruption. On the other hand, the density of the pathway underneath the crater floor did not show a statistically significant change. Moreover, a petrological study indicated that the chemical composition of the 2009 eruption lava ejecta matched the composition of the magma deposit created on the crater floor in the 2004 eruption. From this muographic and geochemical joint analysis, the following scenario for this 2009 eruption has been proposed.[6] Magma did not flow up the pathway in the 2009 eruption, and high-pressure vapor simply blasted through the old magma deposit, which was still strong enough to act as a "plug" of the magma pathway.

The gas flow inside a volcano is generally difficult to directly observe with muography because the expected density variations will be too small to be detected. However, like an example shown above, indirect evidence could be captured as the structural modification as a consequence of phreatic explosions. Needless to say, multi-aspect observations are necessary to interpret this indirect evidence.

2.2.2. *Magma convection*

Satsuma–Iwojima volcano, Japan, almost continuously emits large amounts of magmatic gases without a significant output of magma. Satsuma–Iwojima is located on the rim of Kikai Caldera, a super-volcano caldera that measures 19 km in diameter, and that caused a catastrophic eruption 7,300 years ago with a volcanic eruption index (VEI) of 7. The current magma supply rate to the region beneath Satsuma–Iwojima can be estimated with geodetic measurements, however it will be underestimated if only the expansion rate of the volcano is considered. If the magmatic convection process is assumed to be the correct model for the degassing of the magma, the rate of the magma supply to the shallow part of the volcano will be estimated to be more than one order of magnitude larger than the expansion rate of the volcano.[8] In the magmatic convection model, a magma conduit is connected to a deep magma chamber (Fig. 2), and in the upper part of the conduit, the gas escapes from the magma and exits the volcano. The degassed magma sinks and, concurrently new low-density non-degassed magma ascends from the bottom of the conduit and the cycle continues.

In this convection model of Satsuma–Iwojima, from geochemical studies, it is expected that the degassing pressure of the magma is within a range of 0.5–3.0 MPa[9] because the oversaturation of volatiles in the melt was found, i.e., the degassing had been occurring under relatively low-pressure conditions.[10] The muographic image (Fig. 3) shows that the density gradually decreases up the conduit, and the top of the magma column at 400 m a.s.l. has the lowest density, indicating the presence of magma degassing. The overall features of this first muographic degassing evidence found in Satsuma–Iwojima volcano were consistent with the geochemical prediction from the magmatic convection hypothesis. The convection rate (speed of the magma movements) will be discussed in Section 2.3.

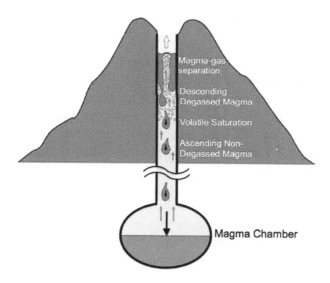

Figure 2. Conduit magmatic convection model of Satsuma–Iwojima volcano, Japan. The red and pink colored regions inside the conduit respectively indicate the non-degassed and degassed magma.

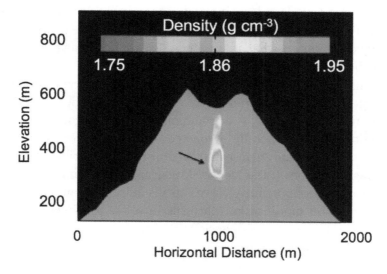

Figure 3. Muographic image of Satsuma–Iwojima volcano, Japan. The arrow (or the green to blue colored region) indicates the location of the bubbly low-density magma.

2.3. Real-time monitoring and multiple exposure muography

The target of the first muographic monitoring of geo-fluids was to image the permeation of rainwater into the mechanical fracture zone of an active seismic fault.[11] The target size was much smaller than volcanoes and thus the measurements were realistic. The results were later applied to monitor the underground water table underneath a landslide fault.[12] Later, this muographic monitor captured magmatic dynamics[13] and tephra fallout[14] in an erupting volcano. In 2013, time-sequential muographic images showed the ascent and descent of the magma head, which synchronized to the visual observation timings of volcanic glows during this eruption episode. The resultant time-sequential muographic images were similar to medical angiographic images. In this case, magma played a role of the contrast agent in angiography. In the field of medical imaging, by taking advantage of recent deep learning techniques, image-processing techniques have been highly developed for automated medical image analysis and evaluation. Angiography-like time-sequential images indicate that this deep learning technique could possibly be applied to the field of muography. The applicability of a deep learning technique similar to what is used now for medical image processing will be introduced in Section 2.6.

As already described in Section 2.2.2, the muographic image of Satsuma–Iwojima suggested that a low-density region at the top of the magma column is a common feature of conduit magma convection. It is generally expected that as water-rich magma ascends to the ground surface and decompresses, volatiles are exolved and magma becomes less dense. Muographic real-time monitoring offers a unique opportunity for us to visualize and analyze magma movements and modification patterns in order to directly test this hypothesis. This picture is consistent with our understanding about the conduit magma convection: the ascending speed of the low-density, vesiculated magma body at the top of the magma column seems faster than that of the higher density magma body.

On June 4, 2013, the eruption alert level had risen from level 1 (signs of volcano unrest) to level 2 (minor eruptive activity) at Satsuma–Iwojima volcano, Japan. For the first time, time-sequential muographic images showed the ascent and descent of the magma head, which synchronized to the timings of volcanic glows during this eruption episode.[13] After the last volcanic glow observation on June 30, muographic images also showed the formation of a new volcanic plug within the volcano. A similar plug formation could also be seen in Sakurajima volcano, Japan.[15] From the ascent and descent speed of magma, researchers could infer that the average magma flow speed near the surface was 30 m day^{-1}.[13] On the other hand, the magma convection rate of Iwojima was estimated to be 10^6 m^3 day^{-1} on the basis of the SO$_2$ emission rate of 550 t day^{-1}.[9] This means, assuming that the conduit has an inner radius of about 200 m, this average flow speed indicates a low-density, vesiculated magma ($<$1 g cm^3) near the surface.

Muographic monitoring also offers an alternative solution to study underground water mobility: (A) variations of bulk water contents over a multi-hectare-scale area and (B) the total mass of water inflow and outflow within a given volume. Risks in landslides and floods in tunnels are all related to the existence and mobility of subterranean water. Usually, the only method to monitor the behavior of underground water is to monitor the subsurface water content with water gauges and soil moisture sensors which are inserted into vertically and horizontally drilled boreholes.[16] However, from only a few data points, it is generally difficult to extrapolate the actual movements of underground water that are affected by complicated surrounding geology. On the other hand, muography measures larger-scale mass movements as a bulk unit.

Tanaka *et al.* (2011)[11] and Tanaka and Sannomiya (2012),[12] respectively, captured muographic images of rainwater indicating a gradual permeation to the underground direction through the vertical fracture zone, and inflation of underground water gradually propagated downstream through the horizontal fracture zone (Fig. 4). These fracture zones were, respectively, associated with seismic faults and landslide faults. As a result, 3 tons/m^2 and 3.6×10^4

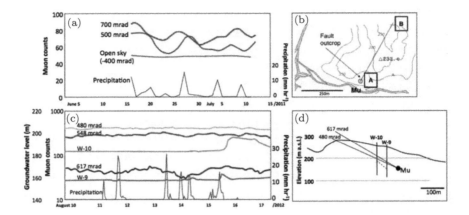

Figure 4. Time-sequential presentation of the response of the muon flux to rainfall events. (a) The response in the Itoigawa–Shizuoka Tectonic Line (ISTL) seismic faults. (b) The geometric configuration of the detector (Mu) and the ISTL seismic faults. The line A-B indicates the estimated near-vertical fault line. (c) The response in the landslide fault, Shizuoka, Japan. (d) The cross-sectional view of the geometric configuration of the detector (Mu) and water gauges in water wells (W-9 and W-11). The dashed lines indicate the ordinary underground water level.

tons of water inflow were, respectively, measured in the vertical and horizontal underground water pathways as a response to the local precipitation of 10^2 mm.

In order to observe dynamics within a shorter time scale than the time resolution of the observation system, time-sequential muographic images can be integrated to study cyclic processes. One of the factors that makes it challenging for us to perform real-time or rapid time-sequence imaging is that the cosmic ray muon flux is relatively low and requires long integration times to reach an adequate contrast for muographic images. However, low cosmic ray muon flux can be compensated for by integrating a large number of short acquisition frames, which is appropriate for periodic processes. After summing up all the images according to its corresponding time point, an average snapshot image with better statistics is obtained.

As a proof-of-concept measurement, a large-scale industrial electric furnace was measured with multiple exposure muography for two electric load patterns.[17] The acquisition frames were triggered by the electric load levels to acquire images for two different patterns: when higher electric load was applied and when lower electric load was applied. When higher electric loads are applied, the temperature rises inside the furnace, then more molten materials are deposited at the bottom of the furnace. The resultant images indicated that when higher electric loads are applied, a thickness of the high-density region increases, and were consistent with our aforementioned vision.

One of the potential applications of multiple exposure muography to Earth science is the investigation of the recurrent cycles of volcanic eruption. Kusagaya (2017)[18] applied multiple exposure muography to Sakurajima volcano, Japan, to detect the magma ascent right before the eruption, and found that the density increases underneath the crater floor correlated with the intervals of the eruption. During the 2009–2017 eruption episode, Sakurajima erupted more than 7,000 times and thus, this volcano was a good target for testing multiple exposure muography.

The technique is applicable to any volcanoes that repeatedly erupt such as Stromboli, for example, which has been erupting every 5 to 20 minutes for thousands of years, along with lava-flows which are emitted several times a year, cascading down on the northern flanks of the volcano.[19] Similarly, the eruptions of Kilauea volcano are characterized by periodic episodes of lava fountaining and the creation of lava lakes.[20] In order for multiple exposure high-speed muography to be feasible, very frequent cycles are necessary. A more challenging multiple exposure muographic measurement would be to image pyroclastic ejection, an event that typically lasts one second after each explosion event; approximately 2,500 eruption events would be necessary to monitor and integrate to detect this event with multiple exposure muography at a 3-sigma confidence level.[a,17]

---

[a]It was assumed that a detector with a detection area of $4\,m^2$ was located at a distance of 200 m from the vent with a radius of 10 m. A uniform rock density of $2\,gcm^{-3}$ was assumed.

2.4.   *From 2D imaging to 3D reconstruction*

In general, volcanoes make good study targets for muographers because they are usually axisymmetric, and it is reasonable to assume that the observed density variations are localized in the vent or crater area. However, in most cases, uncertainty still remains over the shape and alignment of the density anomaly. This uncertainty can be further constrained with bidirectional or multidirectional muographic observations to perform muographic computational axial tomography (Mu-CAT).

Tanaka *et al.* (2010)[21] conducted bidirectional muography to locate a low-density magma pathway found underneath the crater floor in 2006 at Asama volcano, Japan.[6] The shape of the pathway was oval and its size was determined to be 300 ± 100 m in the W–E direction, and 150 ± 50 m in the N–S direction. Furthermore, it was extended toward the north direction. This directional orientation of the pathway explained the reason why Asama volcano has historically ejected pyroclastic and lava flow toward the north direction.

However, there are still several limitations to the present Mu-CAT technique. In bidirectional muography, a grid has to be formed by discretizing a target volume unlike a medical CAT scan that incorporates hundreds of multidirectional images. Therefore, average density in each voxel has to be estimated by providing the researcher-dependent empirical parameters, and there is no way to justify these parameters independently from the measurements. If the detectors can be placed around the target volume more flexibly, the technique used for medical tomography can be applied to the muographic images, however, practical implementation of hundreds of detectors around the target would be challenging in most cases due to the rough terrain and limited infrastructure that are typical around most volcanoes. One possible solution is airborne muography that will be introduced in the next section. Several recent developments include a joint inversion between muographic and gravimetric data.[22, 23]

2.5. *Mobile muography*

2.5.1. *Airborne muography*

Airborne muography has three major advantages over land-based muography: (A) availability of infrastructure required for the measurements (electricity, telecommunication device, etc.), (B) fast transportation and installation of the detector, and (C) optimum position for collection of muon events to approach up to a few meters to the target volume without having to overcome the restrictions of terrain conditions, for example, steep topography (Fig. 5). Muon trackers can be equipped inside the aircraft that flies and hovers near the target of interest to perform muographic measurements. The data

Figure 5.   Photographs of airborne muography. Distant (a) and close-up (b) views of the measurements are shown. The white inset in (a) indicates the region of the close-up view in (b). The distance between the rotor and the cliff is a few meters. Exterior (c) and interior (d) views of the apparatus, an integrated form of an array of detectors and a helicopter, are shown.

can be analyzed in the aircraft in real time to produce a muographic image of the target.

Airborne muography allows us to get the detector as close to the target of interest as possible and, thus, the viewing solid angle (Ω) of the region of interest can be maximized. As a consequence, the recorded number of muons that pass through the region is increased in proportion to Ω that is anti-proportional to a square of the distance between the detector and the target. This is one of the strong points of airborne muography, and is thus far the most effective technique to shorten the time required for capturing muographic images of volcanoes. For example, if the distance to the target is reduced to a half of the original value, it will take a quarter of the original time to produce the same quality images.

There is one more profitable factor for advantage (C) in terms of increasing the contrast of images. It also contributes to a further reduction of the exposure time in the topographically prominent target volume. For example, the diameter of the peak region of the lava dome is typically ~100 m in diameter, but for land-based muography, the path length can exceed more than 2,000 m due to the additional rock from the volcanic edifice.[24] In the case of Mt. Unzen, airborne muography could shorten the time required for imaging the peak region of the lava dome by more than 2 orders of magnitude.[24]

A combination of (A)–(D) will drastically improve the number of multi-directional images to reconstruct a high-quality 3D imaged internal structure map. A few directional muographic images require empirical information, however, airborne muography can produce one muographic image of a lava dome within a few hours.[24] This means that potentially in 10 days of airborne muography measurements, 100 images of the lava dome could be obtained from 100 different directions, and the resultant 3D image would be comparable to the quality of a medical computational axial tomography (CAT) image.

Airborne muography offers an opportunity to obtain a more detailed image of the structure of a target volume, e.g., a lava dome during its growth period, which is difficult to approach during its eruption. Airborne muography conducted by Tanaka (2016)[24] revealed in a few hours that the high-density solid materials (core) were situated inside the low-density brecciated deposits. The time-dependent variations in the core volume fraction provided useful information to us about the dome formation and growth. The thickness (the erosion level) of the talus is one of the key factors, which controls the collapse magnitude since erosion of the talus may expose the hot and dense core, leading to an explosive event.[25] Pyroclastic flows observed during an intense rainfall event in Unzen, Japan, Merapi, Indonesia, Santiaguito, Guatemala, and Mt. Saint Helens, USA, were all caused by the erosion of the talus.[25] Visualization of the internal structure of growing lava domes that could be taken within a few hours would provide a snapshot of the talus thickness and, therefore, repeated measurements would provide direct information on its evolution that might be essential to facilitate risk mitigation prior to volcanic dome collapses.

2.5.2. *Automobile muography*

Automobile muography is another strategy to increase the potential of muography. In particular, this technique enables researchers to map out the subsurface density distribution over a large area ($> 1,000 \, \text{km}^2$) given a sufficient number of tunnels distributed throughout the target area (auto-tunnel muography). For example, Tanaka (2015)[26] collected data from 146 observation points of three different Japanese peninsulas to map out the shallow density distribution of an area measuring $1,340 \, \text{km}^2$ in total, and based on these data from muography, it was subsequently able to compare tectonic differences between these peninsulas (Fig. 6).

Subsurface density variations are usually estimated from a Bouguer gravity anomaly map. However, uncertainties in the Bouguer reduction density (uniform density distribution assumed above the reference surface) affect the estimation of the underground density distribution. In order to reduce these uncertainties, several analysis methods have been proposed. However, an erroneous density distribution can be frequently derived if the target area contains a steep gradient in the Bouguer anomaly and topography (e.g., the area crossing a fault).[27]

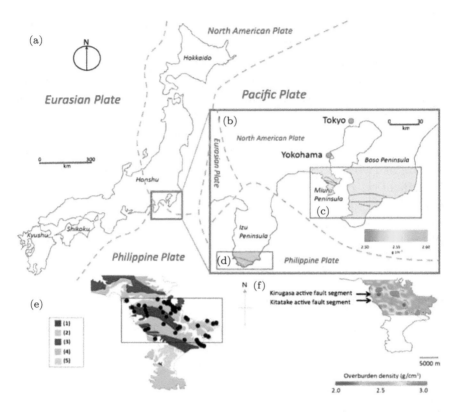

Figure 6. Density variations of the central southern region of Honshu, Japan.
The densities plotted in this map were derived by excluding the data from
the low-density regions located underneath the seismic fault segments. (a) The
map of Japan along with an indication of four tectonic plates. The dashed
lines indicate the interface of these tectonic plates. (b) The enlarged map
showing the geographical configuration of Izu, Miura, and Boso peninsulas.
(c) The muographically surveyed area in Miura and Boso Peninsulas. (d) The
muographically surveyed area in Izu Peninsula. (e) Geological maps of the Miura
peninsula, indicating the distribution of the accretionary complex formed between
40 and 20 million years ago (1), marine and non-marine sediments formed between
15 and 7 million years ago (2), those formed between 7 and 1.7 million years ago
(3), those formed between 1.7 and 0.7 million years ago (4), and those formed
between 0.7 and 0.15 million years ago (5). Black dots show the locations of
observation points. (f) Muographic image of the subterranean density distribution
of Miura peninsula. The colors in this map indicate the terrain density. The region
used for plotting this figure is shown by the inset in (e). Two blue dashed lines,
respectively, indicate Kinugasa and Kitatake active fault segments.

In auto-tunnel muography, the measurements are taken while the vehicle is moving, and the data are generated from integration of the attenuated muon flux over the traveled distance in the tunnel. Using the global positioning system (GPS), the position of the vehicle at the entrance of the tunnel is accurately identified and the acceleration of the vehicle inside a tunnel is measured with a gyro-sensor. The positioning error inside the tunnel is typically less than a few meters. The data collected near both ends of the tunnel (typically 20–30 m from the ends) are discarded in order to remove the erroneous data from overburdens in these areas. Error estimations of the overburden thickness in these areas are usually large because the thickness is comparable to or smaller than the elevation accuracy of the digital elevation map (DEM).

As a result of conducting auto-tunnel muography, density variations were found between the muographic measurement results as obtained in submarine-volcano-originated peninsula tunnels and accretionary prisms-originated peninsula tunnels, which originated from the sediments deposited on the Pacific Ocean floor between 15 and 20 million years ago. The derived subterranean bulk density of these peninsulas was in agreement with gravimetric results.[27] Since volcanic rock is generally more porous, its bulk density is generally less dense than sedimentary rock and thus, can be differentiated from sedimentary rock with muography.

Moreover, Tanaka (2015)[26] compares the average densities determined within the peninsulas with the following geological domains: (A) accretionary complex formed between 40 and 20 million years ago $(2.79 \pm 0.05\,\mathrm{g/cm3})$, (B) marine and non-marine sediments formed between 15 and 1.7 million years ago $(2.57 \pm 0.03\,\mathrm{g/cm}^3)$, (C) marine and non-marine sediments formed between 1.7 and 0.7 million years ago $(2.43 \pm 0.05\,\mathrm{g/cm}^3)$, and (D) basaltic and andesitic rock formed between 7 and 1.7 million years ago $(2.51 \pm 0.05\,\mathrm{g/cm}^3)$. He found a general trend that the density tends to be higher for older rock, and this trend was consistent with gravimetrically determined density values.[27] A closer inspection revealed an interesting density gap between tertiary and quaternary sediments. Tertiary volcanic rocks seemed to have almost the same density as sedimentary rocks of the same age.

One common feature of the muographic maps of these peninsulas with different tectonic origins was the low bulk density along active fault lines. The adjacent low-density points measured along the fault line show relatively higher density than average. Since higher density rocks are likely to be mechanically stronger, such rocks might have terminated further fault developments. Another interesting feature seen in the muographic maps was that lower density values were more often observed in regions near active fault than non-active fault areas. Indications are that the bulk density along the fault lines may correspond to activity levels of the faults since these low-density regions may indicate rock with internal cracks which would be susceptible to frequent fault activities.

For automobile muography measurements, road tunnels are used. Another example of tunnel muography can be seen in the recent developments of glacier muography.[28] If railway or road tunnels are located underneath the targeted glacier, detectors can be installed inside the tunnel so that they can record the number and the angles of muon tracks that arrive after passing through the glacier and the bedrock. This technique was used for modeling the bedrock topography underneath the Eiger glacier, Switzerland, which provided useful information on the bedrock erosion. More detailed description about glacier muography can be found in Chapter 6b: "Application of muography to alpine glaciers." Other interesting examples of recent tunnel muography works include work done inside a British railway tunnel.[29] In this railway tunnel measurement, hidden construction shafts were identified. These shafts could possibly cause the potential collapse of the overburden in the future, and thus muography is expected to be a powerful tool for preventing such accidents.

2.6. *Usage of AI*

Natural disaster forecasting is one of the most critical social demands in modern societies, and researchers have attempted to apply statistical algorithms and machine learning to the earth scientific data to help to meet these social demands.[30–32] In the field of medical image analysis, a deep learning technique has shown its applicability to the automatic image analysis and medical report production. Due to its similarity with X-ray radiography, this technique could be adapted to muography to expand the capabilities of data processing. This could improve the speed and efficiency in analysis of muographic data and contribute to providing more accurate forecasts of volcanic eruptions.

Nomura *et al.* (2020)[33] applied the deep learning technique to muographic data acquired at Sakurajima volcano, Japan, between 2014 and 2016 during the most active period of the last few decades. They investigated the effectiveness of a convolutional neural network (CNN) for eruption forecasting at the Showa crater, of Sakurajima volcano, Japan, and compared this with other methods such as support vector machine (SVM) or neural network (NN) based on daily muographic images (muograms) (Fig. 7).[b]

The results showed moderate performance for day-level eruption forecasting at the active vent of Sakurajima volcano. The AUC (0.726), accuracy (0.714), and specificity (0.857) were highest when

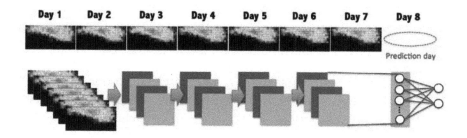

Figure 7. Scheme of machine learning of the muographic images for eruption prediction (top). A time sequential muogram set as input data is used for convolving and connecting layers to generate the final output layers (bottom).

---

[b]464 sets of seven consecutive daily muograms were used. In these sets, 1,439 (Showa crater) and 10 eruptions (Minamidake crater) were included, and the eruption prediction was attempted based on the seven daily muograms prior to the prediction day.

the input to the CNN model was limited to the active vent region. Although it is difficult to know why the CNN model showed the aforementioned result, such a result was obtained because the muon count underneath the active vent region tends to decrease right before the eruption. This is due to the plugging of a magma pathway right before the eruption, which has been already modeled by Iguchi (2013)[34] based on geodetic and seismic measurements.

The work by Nomura *et al.* (2020)[33] utilized the data acquired with the low-definition observation system, and thus the active vent region was limited to 5 × 5 pixels. Currently, a high-definition muography observation system has been under operation since January 2017.[35] Furthermore, the eruption sequence of this volcano has abruptly changed and the next neighbor vent has been activated since November 2017. It is anticipated that a more detailed eruption mechanism model will be clarified by comparing the forecasting performance between these two eruption episodes.

2.7. *Muography art*

As an integral part of human culture, the need for all citizens to understand the recent developments of science is growing. However, due to the complexity of these subjects and the specialized language researchers use, there is a challenge when it comes to science communication between scientists and the communities they serve. Recently, scientists and artists have been collaborating on projects to improve this situation. Pioneering works include programs at CERN, for example art@CERN[36] and art@CMS,[37] which have been facilitating collaborations for this purpose. More recently, muographers have been working with the Muography Art and Muographic Liberal Art Projects.[38] In particular, Fine-Art-Muographers and other international and local organizations have started dialogues between artists, scientists, and the general public about muography within the framework of the ORIGIN Network.[39]

By emphasizing ideas visually with art, the total amount of people that can understand the concepts of science is increased. Muography, with its interdisciplinary characteristics that incorporate topics of particle physics, the universe, earth and planetary sciences, and human cultural heritage such as historic architecture, has the potential to attract a diverse non-scientist audience and encourage science literacy. A high proportion of scientific literacy is becoming increasingly important as scientific topics continue to exert greater influence in our societies. The general public need to understand topics such as the moral/ethical issues of science and the need to continue to fund basic scientific research for the future prosperity and enrichment of our daily lives.

Attracting a diverse spectrum of people to take an interest in science, and specifically muography, can be achieved by communicating more creatively with the cooperation of fine artists. One of the examples can be seen in Fig. 8. Live art performances are particularly engaging and a great way to present physics, which reminds people that natural phenomenon is dynamic, not static. One pioneering work of the Muography Artist in Residence (MAIR) project facilitated a recent Hungarian and Japanese science/art collaboration. A Hungarian ceramic sculptor, Agnes Husz, participated in an MAIR project at Tama Art University, a Japanese art school. Her dynamic installation process was presented in real time as part of the Tama Art University Museum exhibit "Answer from the Universe: Vision Towards the Horizons of Science and Art Through Muography".[40] Agnes Husz's sculpture that combined video, premade sculpture, and sound art with real-time installation art (the clay was shaped, pounded, and stretched in the gallery as audiences watched) became a mirror of the process of muography itself. Another MAIR project generated a piece by an Italian music composer, Federico Iacobucci, titled "Muography Symphony" that was performed at the Italian Institute of Culture in Tokyo in collaboration with Hungarian clarinetist István Kohán and the Tokyo Universal Symphony Orchestra.[41]

These shared experiences of the scientists and artists were in turn shared with audiences at the events. Muography art can be an educational tool that transcends age, gender, and language barriers. Pioneering art education outreach programs were conducted by Kansai University.[43] In particular, a workshop was organized by Fine-Art-Muographers to educate Japanese elementary school

22 H. K. M. Tanaka

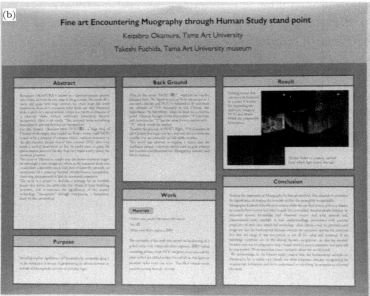

Figure 8. Example of the visual presentation about muography in a form of fine art (a). The duplicated image of HOU, the Chinese mythic bird that is believed to fly 36,000 km high, and muography is presented as an object inseparable from space. A number of drilled holes seen on this folding screen that consists of six $2.5 \times 1.2\,\mathrm{m}^2$ grilled cedar panels indicate both eyeballs and muons. The poster presentation about this artwork done in the style of a scientific research poster is also shown (b).[42]

children and their parents about the topic and helped to ignite the children's interests in science and muography in ways that scientists working alone would not be able to achieve.

3. Perspectives

3.1. *From land to sea*

Due to the unique properties of cosmic-ray muons, muography opens a new possibility to observe kilometric scale (sub-mesoscale) and high frequency (hourly to daily) dynamics that was not possible to monitor with conventional geophysical techniques.

Key to the survival of human civilization and also the source of some of its most dramatic disasters, the processes that affect the shallow seas are of vital importance for researchers to understand in order to better face the challenges of the 21st century. Coastal flooding from extreme storm tide events or tsunami is an example of some hazards that coastal communities face.[44,45] Driven by a combination of astronomical tide and storm surge or seismic tide, the water level changes these events can induce can overtop or breach coastal defense systems and can threaten human lives, property, and infrastructures. Researchers believe that these threats will only increase in the coming century, therefore understanding how these flooding processes interact and combine will be a crucial element that will be necessary to better respond to disasters and try to avoid some of these losses in the future.

A large proportion of the world's largest and most important metropolitan areas are in close proximity to the coastal regions. Tokyo is an example of a highly populated city that is economically reliant on the coastal sea that surrounds it while also being vulnerable to its hazards. At the same time, underneath its coastal sea floor, a huge amount of sustainable energy resource is expected; part of the natural gas reservoir associated with the South Kanto Gas Field (SKGF) exists underneath Tokyo bay. The volume of the biogenic gas deposits in SKGF is thought to occupy more than 90% of the natural gas deposits of the entire nation, and the Tokyo bay area occupies 30% of the total region of SKFG.[46] However, this potential energy

resource underneath Tokyo bay has not been surveyed because most
of the conventional survey techniques require placement of sensors on
the sea surface or on the seafloor, which would interrupt sea traffic
near one of the world's most busy seaports.

One of the largest submarine tunnels in the world, the Trans
Tokyo Bay Aqua Tunnel, which incorporates a bridge and tunnel
of over 25 km, connects the metropolitan areas around the Tokyo
bay. The Aqua Muography project has been established in order to
add a new observation window to deepen our understandings about
the marine dynamics and structure under the seafloor of this very
important coastal sea region, a new detection system is now under
construction, which will be installed in this unique undersea tunnel
space to make the ground floor of the Aqua Tunnel be the Tokyo Bay
Hyper Kilometer Submarine Deep Detector (HKMSDD): a multi-
kilometer-long submarine muographic observation system (Fig. 9).
The HKMSDD is the name of an integrated array of detectors
measuring more than one kilometer and a submarine tube; it is a
detector that in its final form is planned to be 10 km in length for
monitoring Tokyo Bay including its sub-mesoscale ocean dynamics
and the geology of its seafloor.[47]

3.2. *From Earth to outer space*

There have been several attempts to apply muography to extraterrestrial objects such as Mars[48] and Solar System Bodies (SSBs).[49]
Muons can be generated in the planetary atmosphere or in solid
stellar bodies if they don't have their own atmospheres. Although
the energy spectrum of Martian atmospheric muons that can be used
for muography will be completely different from the one available on
Earth, near horizontal muons can be used for muography of Martian
objects.[c] On the other hand, the average muon energy is much lower

---

[c]On Mars, the surface atmospheric pressure is ∼1/100 that on Earth. As a
result, many of the primary particles reach the surface before interaction. This
results in a lower muon production rate than in the Earth's atmosphere. Tanaka
(2007)[50] showed from the steep power law dependence in the early stages of the
shower cascade that the muon flux will be strongly dependent on the atmospheric
pressure, zenith angle, and muon energy. As a result, unlike the horizontal muon

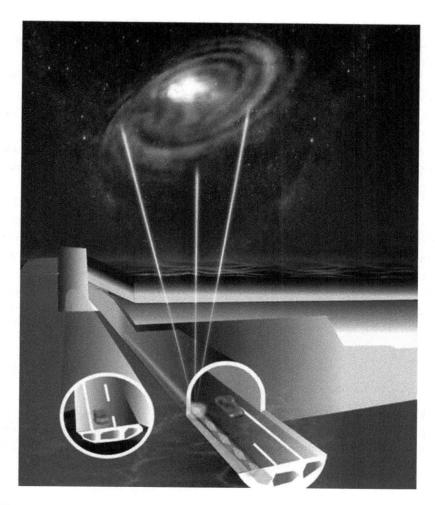

Figure 9. Conceptual view of the Tokyo Bay Hyper Kilometer Submarine Deep Detector.

rate on Earth, the horizontal muon rate is higher than the vertical rate on Mars due to the additional atmospheric mass which the primary cosmic ray must transverse. Considering that the horizontal atmospheric depth of Mars is $100\,\mathrm{g\,cm^{-2}}$, a sufficient number of muons are generated in the horizontal direction, and their energies are sufficiently high because the probability of the secondary meson decay before further interactions is higher than it would be on Earth.

on SSBs than one on Earth or Mars[d]; thus, the potential target size will be more limited in SSB muography. Moreover, up to this date, flight-proven muographic observation systems have not been developed yet.

Silicon trackers or scintillation detectors are a candidate for the detector component of the system designed for outer space muography, but more importantly it is difficult to load the sufficient amount of radiation shield to the space probe or rover for reducing background to detect low-rate muons after passing through the extraterrestrial objects under higher radiation background conditions than the Earth surface.

The time required for measurements, the size (weight) of the detector, and the spatial resolution of the resultant images are critical to consider for outer space muography, and they have a tradeoff relationship. However, for extra-terrestrial muography, the second condition (size and weight) is more strongly restricted, and therefore, it is necessary to find the potential target to be visualized that doesn't require high-resolution images or short-time ($<$ 1year) data gathering and only requires a light detector, but still has the potential to yield useful scientific results.

4. Conclusion

After several attempts to search for hidden structures inside gigantic objects with muography in the 20th century, these trials were proved by displaying the empty magma pathway underneath the solidified

---

[d]Since there is no atmosphere on the small stellar bodies (SSB) such as asteroids or comets, primaries directly collide with the stellar bodies and generate muons inside them. The average energy of these muons is much lower than the atmospheric muons because the secondary mesons are more frequently interacting with the nuclei in the material and, thus, their energies are quickly dispersed and dissipated. The resultant muon flux ($>$ 50 GeV) generated directly in solid rock is 10^3–10^4 times lower than the terrestrial atmospheric muon flux.[49] This flux has a tendency to decrease as a medium density increases because the Meson's mean free path (MFP) is further shortened. The aforementioned flux decreases by one half if the medium density mass is doubled.

magma of an active volcano for the first time early in the 21st century. This evidence clearly showed that muography has a potential to visualize any gigantic object near the scale of or smaller than volcanoes, and muography rapidly entered into a more practical stage. During the last decade, muography has been applied to geological targets, historic architecture, and social infrastructure and the technical principle has been demonstrated by displaying density anomalies inside these targets. However, the technique is still far from its full potential. After a proof of concept was obtained in the field measurements, it has been necessary to push it forward technically to offer more practical, beneficial, and profitable solutions for researchers and engineers in related fields. To achieve this, concretized cooperation and collaboration with researchers in different fields will be profitable. Cooperation between science and humanities experts is one strategy to extend the capabilities of muography. One example would be to investigate the remnants of past disasters recorded inside the historic objects; in order to do this successfully we need to know (A) the anticipated physical properties of the remnants that help us to estimate the magnitude of the disaster and (B) when exactly this disaster occurred. If the remnant is recorded inside gigantic objects, muography can contribute to factor (A), and historians who studied this historic object can contribute to factor (B). This kind of collaboration took place in 2019. A collaboration between muography researchers, historians, and engineers specializing in disaster prediction revealed physical evidence of a large crack generated by a major earthquake (M 7.25–7.75 on September 5, 1596) in an ancient mound in Japan.[51] Based on muographic data of the alignment and size of the crack, the mound collapse was modeled, and the robustness of the bedrock was discussed in order to better understand this past disaster and to ascertain how this new information can be utilized to improve preparation for future earthquakes. I foresee that muographic research in the next decade will be dedicated toward this direction of expanding innovation and collaboration.

References

1. E.P. George, Cosmic rays measure overburden of tunnel, *Commonwealth Engineering*, **1955**, 455–457, (1955).
2. L.W. Alvarez *et al.* Search for hidden chambers in the pyramid, *Science*, **167**, 833–839, (1970).
3. K. Nagamine, M. Iwasaki, K. Shimomura, and K. Ishida, Method of probing inner-structure of geophysical substance with the horizontal cosmic-ray muons and possible application to volcanic eruption prediction, *Nuclear Instruments and Methods in Physics Research A*, **356**, 585–595, (1995).
4. E. Caffau, F. Coren, and G. Giannini, Underground cosmic-ray measurement for morphological reconstruction of the Grotta Gigante natural cave, *Nuclear Instruments and Methods in Physics Research A*, **385**, 480–488, (1997).
5. H.K.M. Tanaka, T. Uchida, M. Tanaka, M. Takeo, J. Oikawa, T. Ohminato, Y. Aoki, E. Koyama, and H. Tsuji, Detecting a mass change inside a volcano by cosmic-ray muon radiography (muography): First results from measurements at Asama volcano, *Japan, Geophysical Research Letter*, **36**, L17302, (2009a). doi:10.1029/2009GL039448.
6. H.K.M. Tanaka, *et al.* High resolution imaging in the inhomogeneous crust with cosmic-ray muon radiography: The density structure below the volcanic crater floor of Mt. Asama, *Japan, Earth and Planetary Science Letters*, **263**, 104–113, (2007a).
7. B. Urabe, N. Watanabe, and M. Murakami, Topographic change of the summit crater of Asama Volcano during the 2004 eruption derived from airborne synthetic aperture radar (SAR) measurements, *Bulletin of Geographical Survey Institute*, **53**, 1–6, (2006).
8. R. Kazahaya, and T. Mori, Interpretations for magmatic process and eruptive phenomena by way of volcanic gas studies, *Bulletin of the Volcanological Society of Japan*, **61**, 155–170, (2016).
9. K. Kazahaya, H. Shinohara, and G. Saito, Degassing process of Satsuma–Iwojima volcano, Japan: Supply of volatile components from a deep magma chamber, *Earth Planets and Space*, **54**, 327–335, (2002).
10. J.W. Hedenquist, M. Aoki, and H . Shinohara, Flux of volatiles and ore-forming metals from the magmatic-hydrothermal system of Satsuma Iwojima volcano, *Geology*, **22**(7), 585–588 (1994).
11. H.K.M. Tanaka, *et al.* Cosmic muon imaging of hidden seismic fault zones: Raineater permeation into the mechanical fracture zone in Itoigawa-Shizuoka Tectonic Line, *Japan, Earth and Planetary Science Letters*, **306**, 156–162, (2011).

12. H.K.M. Tanaka, and A. Sannomiya, Development and operation of a muon detection system under extremely high humidity environment for monitoring underground water table, *Geoscientific Instrumentation Methods and Data Systems*, **2**, 719–736, (2012).

13. H.K.M. Tanaka, T. Kusagaya, and H. Shinohara, Radiographic visualization of magma dynamics in an erupting volcano, *Nature Communication*, **5**, 3381, (2014). doi:10.1038/ncomms4381.

14. H.K.M. Tanaka, Development of the muographic tephra deposit monitoring system, *Scientific Reports*, **10**, 14820, (2020). https://doi.org/10.1038/s41598-020-71902-1

15. L. Olah, H.K.M. Tanaka, T. Ohminato, G. Hamar, D. Varga, Plug formation imaged beneath the active craters of Sakurajima volcano with muography, *Geophysical Research Letter*, **46**, (2019). doi:10.1029/2019GL084784.

16. E. Yuliza, H. Habil, M.M. Munir, M. Irsyam, M. Abdullah, and Khairurrijal, Study of soil moisture sensor for landslide early warning system: Experiment in laboratory scale, *Journal of Physics: Conference Series*, **739**, 012034, (2015). doi:10.1088/1742-6596/739/1/012034.

17. H.K.M. Tanaka, Development of stroboscopic muography, *Geoscientific Instrumentation Methods and Data Systems*, **2**, 41–45, (2013). doi:10.5194/gi- 2-41-2013.

18. T. Kusagaya, Reduction of the background noise in muographic images for detecting magma dynamics in an active volcano. Ph.D. Thesis, The University of Tokyo (2017). Retrieved from https://repository.dl.itc.u-tokyo.ac.jp/?action=pages_view_main &active_action=repository_view_main_item_detail&item_id=52347& item_no =1&page_id=28&block_id=31

19. L. D'Auria,: Perspectives of muon radiography for Stromboli, International Workshop on Muon Radiography of Volcanoes, 11– 12 October 2010, Naples, Italy, 2010.

20. M.O. Garcia, A.J. Pietruszka, J.M. Rhodes, and K. Swanson, Magmatic processes during the prolonged Pu'u 'O'o eruption of Kilauea Volcano, Hawaii, *Journal of Petroleum Science and Engineering*, 967–990, (2000). doi:10.1093/petrology/41.7.967, 2000.

21. H.K.M. Tanaka,*et al.* Three dimensional CAT scan of a volcano with cosmic ray muon radiography, *Journal of Geophysical Research*, **115**, B12332, (2010). doi:10.1029/2010JB007677.

22. R. Nishiyama, Y. Tanaka, S. Okubo, H. Oshima, H.K.M. Tanaka, and T. Maekawa, Integrated processing of muon radiography and gravity anomaly data toward the realization of high-resolution 3D density structural analysis of volcanoes: Case study of Showa-Shinzan

lava dome, Usu, Japan, *Journal of Geophysical Research Solid Earth*, (2014). doi:org/10.1002/2013JB010234.

23. A. Barnoud, V. Cayol, V. Niess, C. Cârloganu, P. Lelièvre, P. Labazuy, and E.L. Ménédeu, Bayesian joint muographic and gravimetric inversion applied to volcanoes, *Geophysical Journal International*, **218**, 2179–2194, (2019). doi:org/10.1093/gji/ggz300.

24. H.K.M. Tanaka, Instant snapshot of the internal structure of Unzen lava dome, *Japan with airborne muography*. *Scientific Reports*, **6**, 39741, (2016). doi:10.1038/srep39741.

25. A.J. Hale, Lava dome growth and evolution with an independently deformable talus, *Geophysical Journal International*, **174**, 391–417, (2008).

26. H.K.M. Tanaka, Muographic mapping of the subsurface density structures in Miura, Boso and Izu peninsulas, *Japan. Scientific Reports*, **5**, 8305, (2015). doi:10.1038/srep08305.

27. K. Nawa, Y. Fukao, R. Shichi, and Y. Murata, Inversion of gravity data to determine the terrain density distribution in southwest Japan, *Journal of Geophysical Research*, **102**, 27703–27719, (1997).

28. R. Nishiyama, A. Ariga, T. Ariga, *et al.* Bedrock sculpting under an active alpine glacier revealed from cosmic-ray muon radiography, *Scientific Reports*, **9**, 6970, (2019). https://doi.org/10.1038/s41598-019-43527-6

29. L.F. Thompson, J.P. Stowell, S.J. Fargher, C.A. Steer, K.L. Loughney, E.M. O'Sullivan, J.G. Gluyas, S.W. Blaney, and R.J. Pidcock, Muon tomography for railway tunnel imaging, *Physical Review Research*, **2**, 023017, (2020).

30. A. Brancato, *et al.* K-CM application for supervised pattern recognition at Mt. Etna: An innovative tool to forecast flank eruptive activity, *Bulletin of Volcanology*, **81**, 40, (2019). https://doi.org/10.1007/s00445-019-1299-4

31. M. Rüttgers, S. Lee, S. Jeon, and D. You, Prediction of a typhoon track using a generative adversarial network and satellite images, *Scientific Reports*, **9**, 6057, (2019). doi:org/10.1038/s41598-019-42339-y.

32. A. Mostajabi, D.L. Finney, M. Rubinstein, and F. Rachidi, Nowcasting lightning occurrence from commonly available meteorological parameters using machine learning techniques, *Climate and Atmospheric Science*, **2**, 41, (2019). doi:org/10.1038/s41612-019-0098-0.

33. Y. Nomura, M. Nemoto, N. Hayashi, S. Hanaoka, M. Murata, T. Yoshikawa, Y. Masutani, E. Maeda, O. Abe, and H.K.M. Tanaka, Pilot study of eruption forecasting with muography using convolutional neural network, *Scientific Reports*, **10**, 5272, (2020). doi:10.1038/s41598-020-62342-y.

34. M. Iguchi, Magma movement from the deep to shallow Sakurajima Volcano as revealed by geophysical observations, *Bulletin of Volcanology Society Japan.*, **58**, 1–18, (2013).
35. L. Olah, H.K.M. Tanaka, T. Ohminato, and D. Varga, High-definition and low-noise muography of the Sakurajima volcano with gaseous tracking detectors, *Scientific Reports*, **8**, 3207, (2018). doi:10.1038/s41598-018-21423-9.
36. CERN CERN project brings science and art together, CERN Courier, 41, 23, (2001). Retrieved from http://cds.cern.ch/record/1733190/files/vol41-issue4-p023-e.pdf?version=1
37. M. Hoch, and A. Alexopoulos, Art@ CMS and Science&Art@ School: Novel education and communication channels for particle physics, *Astroparticle, Particle, Space Physics, Radiation Interaction, Detectors and Medical Physics Applications*, **8**, 728–736, (2014). doi:org/10.1142/9789814603164_0115.
38. H.K.M. Tanaka, and L. Oláh, Overview of muographers, *Philosophical Transactions of the Royal Society A*, **377**, (2019). doi:org/10.1098/rsta.2018.0143.
39. Muographix ORIGIN Exhibit Tour of Southeast Europe (2018). Retrieved from https://news.muographix.u-tokyo.ac.jp/en/2018/09/19/origin-exhibit-tour-of-southeast-europe/news-local/
40. Tama Art University Museum (2018). Exhibit "Answer From the Universe: Vision Towards the Horizons of Science and Art Through Muography" 19 May–17 June 2018. Retrieved from http://www.tama bi.ac.jp/museum/exhibition.htm.
41. Ministero degli Affari Esteri Catalago Totale Definitivo, GIAPPONE, 363, (2019). Retrieved from https://www.esteri.it/mae/resource/doc/2019/03/catalogo%20totale%20definitivo-291-560.pdf
42. Okamura, and Fuchida, Fine art encountering muography through human study stand point, *Muographers*, 2018, 28 November–1 December, 2018, Tokyo, Japan.
43. Kansai University (2017). Muography Art Project. Retrieved from https://wps.itc.kansai-u.ac.jp/ku-map/
44. K. Guihou, J. Polton, J. Harle, S. Wakelin, E. O'Dea, and J. Holt, Kilometric scale modeling of the North West European Shelf Seas: Exploring the spatial and temporal variability of internal tides, *Journal of Geophysical Research*, (2017). doi:org/10.1002/2017JC012960.
45. M.J. Lewis, T. Palmer, R. Hashemi, P. Robins, A. Saulter, J. Brown, H. Lewis, and S. Neill, Wave-tide interaction modulates nearshore wave height, *Ocean Dynamics*, **69**, 367–384, (2019).

46. J. Matsushima, Biogenic Gas in Tokyo Bay and Bessi Copper Mine, MAGMA Workshop, Muographers 2019, Muographers, September 26, 2019, Tokyo.

47. H.K.M. Tanaka, J. Matsushima, L. Oláh, D. Varga, L. Thompson, J. Gluyas, D.L. Presti, T. Kin, K. Shimazoe, C. Ferlito, G. Leone, N. Hayashi, T. Ohminato, H. Takahashi, H. Mori, H. Miyamoto, S. Abe, K. Akama, J. Ando, R. Aoyagi, E. Asakawa, S.J. Balogh, D.L. Bonanno, G. Bonanno, C. Bozza, R.C. Gamboa, O. Fujimoto, G. Gallo, G. Galgóczi, Á.L. Gera, G. Hamar, S. Hanaoka, T. Hayashi, A. Homma, H. Ichihara, A. Iida, T. Iizuka, T. Imazumi, A. Ishii, A. Kamimura, K. Kamura, O. Kamoshida, S. Kanazawa, K. Kashihara, N. Kawai, S. Kawasaki, S. Kawase, Y. Kobayashi, T. Kumagai, S. Kuroda, T. Kusagaya, T. Koike, E. Maeda, R.A.M. Vásquez, Y. Masutani, K. Miyakawa, S.-i. Miyamoto, K. Miyazawa, I. Mizukoshi, M. Murata, A. Musumarra, M. Nemoto, T. Nibe, T. Nishiki, E. Nishyama, K. Noda, Y. Nomura, G. Nyitrai, Y. Oda, T. Ohara, T. Ohmura, M. Onishi, Y. Ohnishi, H. Ohnuma, M.G. Pellegriti, F. Riggi, G. Romeo, P.L. Rocca, T. Shibata, N. Shimizu, D. Shimokawa, C. Steer, P. Stowell, H. Suenaga, S. Sugimoto, K. Sumiya, K. Suzuki, Y. Tabuchi, K. Takai, T. Ueda, K. Watanabe, T. Watanabe, Y. Watanabe, K. Yamaoka, T. Yokota, and T. Yoshikawa, *Letter of Intent Multi-Aspect Geo-Muography Array (MAGMA) Experiment*, 1–84, (2019).

48. S. Kedar, H.K.M. Tanaka, C.J. Naudet, C.E. Jones, J.P. Plaut, and F.H. Webb, Muon radiography for exploration of Mars geology, *Geoscientific Instrumentation Methods and Data Systems*, **2**, 157–164, (2013).

49. T.H. Prettyman, S.L. Koontz, L.S. Pinsky, A.M. Empl, D.W. Ittlefehldt, B.D. Reddell, M.V. Sykes, and NIAC Phase I Final Report, The National Aeronautics and Space Administration (2013) *Deep Mapping of Small Solar System Bodies with Galactic Cosimic Ray Secondary Showers* (Grant Number NNX13AQ94G) Retrieved from https://www.nasa.gov/sites/default/files/files/Prettyman_2013_PhI_MuonDeepMapping.pdf

50. H.K.M. Tanaka, Monte-Carlo simulations of atmospheric muon production: Implication of the past Martian environment, *Icarus*, **191**, 603–615, (2007).

51. H.K.M. Tanaka, K. Sumiya, and L. Oláh, Muography as a new tool to study the historic earthquakes recorded in ancient burial mounds, *Geoscientific Instrumentation Methods and Data Systems*, **9**, 357–364, (2020).

https://doi.org/10.1142/9789811264917_0002

Chapter 2

Cosmic Ray Muons

Peter K. F. Grieder

*Physikalisches Institut, University of Bern, Sidlerstrasse 5,
3012 Bern, Switzerland*

1. Introduction

The cosmic ray muon flux in the atmosphere, underground, or under water is the result of the hadronic interactions of the energetic component of the primary cosmic radiation with the nucleons and nuclei of the atmospheric constituents as it propagates in the atmosphere. These interactions produce energetic secondary particles, predominantly pions ($\pi^\pm$, π^0), kaons ($k^\pm$, k^0_S, k^0_L), and more rarely charmed particles (D-mesons, J/ψ, etc.) as well as other unstable particles that may interact or decay, in some cases through a chain of decays, as they continue to penetrate deeper into the atmosphere. The end products of this cascading process are atmospheric muons ($\mu^\pm$). These are themselves subject to decay but have a much longer mean life than their parents, which is 2.197 μs at rest, that is extended by the Lorentz factor for energetic muons, or they may interact.

The flux of the cosmic radiation in the atmosphere, including the flux of the secondary component, increases with increasing atmospheric depth and reaches a maximum at an altitude of about 20 km, called the Pfotzer maximum.[1,2] From there on the flux declines about exponentially down to sea level. Below an atmospheric depth of about $600\,\text{g/cm}^2$, muons become the dominating component of the cosmic radiation in the atmosphere.

2. The Primary Cosmic Radiation

The primary cosmic radiation consists predominantly of protons, alpha particles, and heavier nuclei, i.e., of electrically charged hadrons. It is essentially isotropic and of galactic origin, except for the rare UltraHigh Energy (UHE) component ($E \geq 10^{16} - 10^{17}\,\mathrm{eV}$) which is of extragalactic origin.[a] The flux of the incident gamma radiation at the top of the atmosphere is much lower than that of the charged hadrons and plays an insignificant role in the context of this book.

Since the primary radiation consists of charged particles, it is subject to magnetic deflection when propagating in magnetic fields. Thus, it is affected by the galactic, the interplanetary, the magnetospheric, and the geomagnetic magnetic fields while in transit from their source to Earth, and while propagating in the atmosphere.

The galactic magnetic field strength varies from about 6 μG (0.6 nT) near the Sun to 20–40 μG (2–4 nT) in the galactic center region. It amounts to about $10\,\mu$G (1 nT) at a radius of $\sim$3kpc from the galactic center.[3] The interplanetary magnetic field (IMF) strength is about 50 μG (5 nT) at the Earth's orbit. The magnetospheric field is the sum of the current fields within the space bound by the magnetopause and is subject to significant variability, while the geomagnetic field is generated by sources inside the Earth and undergoes secular changes. The combined fields are typically 0.3–0.6 G (30–60 μT) at the Earth's surface, depending on the location. Time-dependent variations are on the order of 1%. The electrically charged secondary cosmic ray component produced in the atmosphere is also subject to geomagnetic effects, as well as the charged decay products.

The absolute flux of the incident primary cosmic radiation on top of the atmosphere varies slightly from location to location on Earth. This is because the local magnetic field strength varies slightly from location to location. Consequently, the secondary cosmic radiation in the atmosphere reflects this situation across the atmosphere when

---

[a]Note that solar particles do not belong in the category of cosmic rays.

comparing intensities at the same altitude. Moreover, the local atmospheric conditions are not the same and affect the particle flux.

The lower energy cosmic ray flux is modulated by the solar activity because of the varying solar magnetic field. It manifests the 11-year solar cycle and for certain effects also a 22-year cycle, and other solar influences. It should be remembered that the solar magnetic dipole reverses polarity at every solar maximum, thus imposing a 22-year cycle as well. Both cycles cause various effects on the galactic cosmic radiation in the heliosphere. The modulation effects decrease with increasing energy and become insignificant for particles with rigidities in excess of ~10 GV.

3. Propagation of the Cosmic Radiation in the Atmosphere

Upon entering the atmosphere the primary cosmic radiation is subject to interactions with the electrons, nucleons, and nuclei of atoms and molecules that constitute the air. As a result, the composition of the radiation changes because of fragmentation as it propagates through the atmosphere. All particles, including the secondaries, suffer energy losses through hadronic and/or electromagnetic processes.

Above an energy of a few GeV, local penetrating particle showers are produced, resulting from the creation of mesons and other secondary particles in the collisions. Energetic primaries, and in case of heavy primaries their spallation fragments, continue to propagate in the atmosphere and interact successively, producing more particles along their trajectories, and likewise for the newly created energetic secondaries. The most abundant particles emerging from energetic hadronic collisions are pions, but kaons, hyperons, charmed particles, and nucleon–antinucleon pairs are also produced.

Energetic primary protons undergo on average 12 interactions along a vertical trajectory through the atmosphere down to sea level, corresponding to an *interaction mean free path* (i.m.f.p.), λ_i [g/cm^2], of about 80 g/cm^2. Thus, a hadron cascade is frequently created which is the parent process of an *extensive air shower* (EAS).[4–10]

In energetic collisions, atmospheric target nuclei get highly excited and evaporate light nuclei, mainly alpha particles and nucleons of energy $\leq 15\,\text{MeV}$ in their rest frame.

The majority of the heavy nuclei of the primary radiation are fragmented in the first interaction that occurs at a higher altitude than for protons because of the much larger *interaction cross-section*, σ_i [cm^2], and correspondingly shorter interaction mean free path, λ_i. The following expression describes the relation between the cross-section and interaction mean free path.

$$\lambda_i = \frac{A}{N_A\,\sigma_i}\ [\text{g/cm}^2],\tag{1}$$

where N_A is *Avogadro's number* ($6.02 \cdot 10^{23}$), A, the *mass number* of the target nucleus, and σ_i, the cross-section for the particular interaction. For a projectile nucleus with mass number $A = 25$, the interaction mean free path is approximately $23\,\text{g/cm}^2$ in air. Hence, there is practically no chance for a heavy nucleus to penetrate down to sea level.

The proton–nucleus (or more general the nucleon–nucleus) interaction cross-section, $\sigma_{p,A}$, scales with respect to the proton–proton (or nucleon–nucleon) cross-section, $\sigma_{p,p}$, approximately as

$$\sigma_{p,A}(E) = \sigma_{p,p}(E)\,A^\alpha.\tag{2}$$

For nucleon projectiles, $\alpha = 2/3$ and $\sigma_{p,p}(E)$ varies slowly over a range of many decades in energy from ~40 mb at 10 GeV to ~80 mb at 10^7 GeV. The energy dependence of the total and elastic proton–proton (pp), antiproton–proton ($\bar{p}p$), and pion–proton (π^+p, π^-p) cross-sections as a function of energy is illustrated in Fig. 1.[11, 12] For details of $\bar{p}p$ cross-sections, see Abe *et al.* (1994).[13] The high energy data were obtained with $\bar{p}p$ colliders. For pion projectiles, $\alpha = 0.75$ and $\sigma_{\pi,p}$ is approximately 26 mb. Note that the relevant cross-section for particle production by the cosmic radiation in the atmosphere is the *inelastic cross-section*.[b]

---

[b]The inelastic cross-section is the difference between the total and the elastic cross-sections. In elastic collisions (elastic scattering), the initial and the final particles are the same, whereas in inelastic collisions they are different, and particle production may occur if energetically allowed.

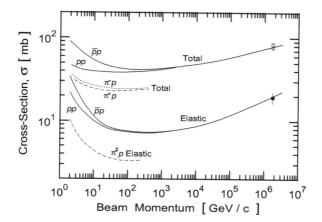

Figure 1. Total and elastic cross-sections for proton–proton (pp), antiproton–proton $(\bar{p}p)$, and pion–proton $(\pi^{\pm}p)$ collisions as a function of beam momentum in the laboratory frame of reference.[11,12]

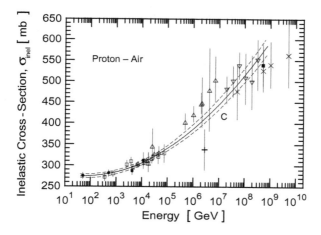

Figure 2. Compilation of inelastic cross-sections for proton–air interactions as a function of energy.[14,15] The symbols refer to the following references: •, C Mielke *et al.* (1994);[14] □, Yodh *et al.* (1983);[16] △, Gaisser *et al.* (1987);[17] ▽, Honda *et al.* (1993);[18] ■, Baltrusaitis *et al.* (1984);[19] +, Aglietta *et al.* (1999);[20] ×, Kalmykov *et al.* (1997).[21]

A compilation of data from measurements of the energy dependence of the inelastic cross-section for proton–air interactions is illustrated in Fig. 2.[14] The data are from a wide variety of experiments, including very indirect cross-section determinations from air shower studies at the higher energies. The solid curve, C, plotted in this figure represents a fit of the form

$$\sigma_{inel} = c + a\log(E) + \log^2(E) \text{ [mb]}, \tag{3}$$

where the parameters have the values $a = -8.7 \pm 0.5$ [mb], $b = 1.14 \pm 0.05$ [mb], $c = 290 \pm 5$ [mb], and E is in [GeV].

4. Secondary Particles

High energy strong interactions as well as electromagnetic processes, such as pair production, lead to the production of secondary particles. Charged pions as well as the less abundant kaons, other mesons, hyperons, and nucleon–antinucleon pairs emerging from strong interactions of energetic primaries with atmospheric target nuclei continue to propagate and contribute to the flux of hadrons in the atmosphere. Of all the secondaries, pions (π^+, π^-, π^0) are the most abundant.

If sufficiently energetic secondary hadrons will themselves initiate new hadronic interactions, they will produce secondaries and build up a hadron cascade that forms the backbone of extensive air showers. However, unstable particles such as pions, kaons, and other particles are also subject to decay.

The competition between interaction and decay depends on the mean life and energy of the particles and on the density of the medium in which they propagate. For a given particle propagating in the atmosphere, the respective probabilities for the two processes become a function of energy, altitude, and zenith angle.

Due to a very short mean life ($\tau \simeq 10^{-16}$ s), neutral pions decay almost instantly into two photons, contributing subsequently to electromagnetic channels.

The energy dependence of the average number of charged secondaries, i.e., the charged particle *multiplicity*, $< n^\pm >$, emerging from a high energy nucleon–nucleon collision, can be described with either of the following relations.[22]

$$< n^\pm >= a + b\ln(s) + c(\ln(s))^2, \tag{4}$$

Table 1. Parameters for energy-multiplicity relation in center of mass energy range 3 GeV–546 GeV.[22]

Equation	a	b	c
4	0.98 ± 0.05	0.38 ± 0.03	0.124 ± 0.003
5	-4.2 ± 0.21	4.69 ± 0.18	0.155 ± 0.003

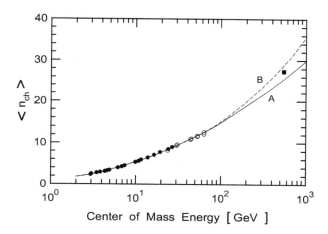

Figure 3.   Mean charged multiplicity of inelastic pp or $\bar{p}p$ interactions as a function of center of mass energy, $\sqrt{s}$. The solid curve, A, is a fit to Eq. 4, the dashed curve, B, to Eq. (5) (after Ref. 22).

or

$$\langle n^{\pm} \rangle = a + bs^c, \qquad (5)$$

where s is the center of mass energy squared. For all inelastic processes, Alner *et al.* (1987)[22] specify for the constants a, b, and c the values listed in Table 1.

The first term in Eqs. (4) and (5) represents the diffraction, fragmentation, or isobar part, depending on the terminology used. The second and third terms, if applicable, account for the bulk of particles at high collision energies that result mostly from central processes. Figure 3 shows the center of mass energy dependence of the secondary particle multiplicity in proton–proton and proton–antiproton collisions obtained from experiments performed at CERN with the proton synchrotron (PS), the intersecting storage ring (ISR), and the proton–antiproton ($\overline{p}p$) collider (UA5 experiment[22]). Relevant for particle production is the energy which is available in the center of mass. At high energies, pp and $\overline{p}p$ interactions behave alike (see Fig. 1). The solid and dashed curves, A and B, are fits using Eqs. (4) and (5), respectively, with the parameters listed in Table 1.

An earlier study of the energy–multiplicity relation that includes mostly data from cosmic ray emulsion stack and emulsion chamber experiments at energies up to 10^7 GeV in the laboratory frame was made by Grieder (1972 and 1977).[23, 24] This author found the same basic energy dependence as given in Eq. (5) but with a larger value for the exponent c. Part of the reason for the higher value of c resulting from this work is probably the fact that the analysis included mostly nucleon–nucleus collisions, which yield higher multiplicities than nucleon–nucleon collisions at comparative energies. Nucleus–nucleus collisions yield even higher multiplicities, but they can easily be distinguished from nucleon–nucleus collisions when inspecting the track density of the incident particle in the emulsion.

5.   Muon Production

5.1.   *Main channels*

Muons (μ^+, μ^-) are the decay products of unstable secondary particles emerging from high energy hadronic collisions. They are copiously produced in hadron cascades, and in air showers, where they account for about 10% of the total particle flux in an average shower at ground level. Their mass is 105.66 MeV and the mean life at rest is $2.197 \cdot 10^{-6}$ s. They decay into an electron ($e^\pm$), a muon neutrino ($\overset{(-)}{\nu_\mu}$), and an electron neutrino ($\overset{(-)}{\nu_e}$), as shown in reactions (6) and (9). The main contributors to the muon component are charged pions (π^+, π^-) and, to a lesser extent, kaons (K^+, K^-, K_L^0), but also charmed particles, such as $D^\pm$, D^0, J/ψ, and others. Apart from photons and neutrinos, muons are the most abundant particles in the cosmic radiation in the lower atmosphere, above an energy of ~1 GeV. Their fraction increases with increasing atmospheric depth down to sea level.

For the basic production reaction,

$$p + p \to p + n + \pi^+$$
$$\pi^+ \to \mu^+ + \nu_\mu$$
$$\mu^+ \to e^+ + \nu_e + \overline{\nu_\mu}, \tag{6}$$

the threshold energy in the laboratory frame of the projectile proton, $E_{p,th,lab}$, in the fixed target proton–proton interaction needed to produce a single π^+ is readily obtained from the following kinematic relation:

$$s = 2E_{p1}E_{p2} + m_{p1}^2 + m_{p2}^2 = m_{p1}^2 + m_{p2}^2 + m_\pi^2, \tag{7}$$

where s is the center of mass energy squared, E_{p1}, m_{p1} and E_{p2}, m_{p2} are the projectile and target proton energies and masses, and m_π is the pion mass. From this relation, we get for the threshold energy of the proton projectile the following equation:

$$E_{p,th,lab} \geq m_p \left(1 + \frac{m_\pi^2 + 4m_\pi m_p}{2m_p^2} \right). \tag{8}$$

The resulting kinetic threshold energy, T, of the proton needed to trigger the reaction is $T \geq E_{p,th,lab} - m_p$, which amounts to $\geq 280\,\text{MeV}$.

The cross-section for this process is $\sigma_{pp} \simeq 4 \cdot 10^{-26}\,\text{cm}^2$ near threshold and rises slowly with proton energy to $\simeq 10^{-25}\,\text{cm}^2$ at about $10^6\,\text{GeV}$.

The following reaction is a typical example of the dominating muon production channel:[c]

$$p + N \text{ (or } A) \to p + N \text{ (or } A') + n\pi^{\pm,0} + X$$
$$\pi^\pm \to \mu^\pm + {}^{(\overline{\nu_\mu})}$$
$$\mu^\pm \to e^\pm + {}^{(\overline{\nu_e})} + {}^{(\overline{\nu_\mu})}, \tag{9}$$

where X stands for anything, including additional hadrons. For incident nuclei, the analogous reaction takes place, producing muons and neutrinos.

---

[c]For further details concerning decay channels of the different unstable particles, the interested reader is referred to the "Review of Particle Physics", assembled by the Particle Data Group; Olive *et al.* (2014).[25]

The bulk of the decay processes that yield muons are two-body decays with a *muon–neutrino* (or *muon–antineutrino*) associated, to satisfy the conservation laws. Consequently, a significant neutrino component is co-produced and continuously building up in a hadronic cascade or an air shower, together with the muon component as the shower propagates in the atmosphere and develops in age.

The mean life of charmed particles is $\leq 10^{-12}$ s. Hence, charmed particle decays are *prompt decays* and yield so-called *prompt or direct muons* that are in general energetic for kinematic reasons. All long-lived unstable particles ($10^{-8} \leq \tau \leq 10^{-10}$ s) are subject to competition between interaction and decay as they propagate in the atmosphere. The probability for either process to occur depends on the mean life of the particle and is a function of its kinetic energy and the local atmospheric density, which is a function of altitude, and therefore also a function of the inclination of the trajectory in the atmosphere, i.e., of the zenith angle, θ.

This interrelationship is responsible for the zenith angle enhancement of the bulk of the muons in air showers, a phenomenon which muons from charmed particle decays do not exhibit. In Table 2, we list the particles and decay schemes that are the most relevant contributors to muons resulting from high energy interactions.

Table 2. Major muon and muon-neutrino parent particles and decay schemes.

Particle symbol	Partial decay modes	Branching fraction (%)	Mean life [s]
$\pi^{\pm}$	$\rightarrow \mu^{\pm} + {}^{(}\overline{\nu_{\mu}}{}^{)}$	99.99	$2.603 \cdot 10^{-8}$
$K^{\pm}$	$\rightarrow \mu^{\pm} + {}^{(}\overline{\nu_{\mu}}{}^{)}$	63.43	$1.238 \cdot 10^{-8}$
	$\rightarrow \pi^{0} + \mu^{\pm} + {}^{(}\overline{\nu_{\mu}}{}^{)}$	3.27	
$\tau^{\pm}$	$\rightarrow \mu^{\pm} + {}^{(}\overline{\nu_{\mu}}{}^{)} + {}^{(}\overline{\nu_{\tau}}{}^{)}$	17.36	$2.906 \cdot 10^{-13}$
$D^{\pm}$	$\rightarrow {}^{(}\overline{K^{0}}{}^{)} + \mu^{\pm} + {}^{(}\overline{\nu_{\mu}}{}^{)}$	7.0	$1.040 \cdot 10^{-12}$
D^{0}	$\rightarrow \mu^{+} + Hadrons$	6.5	$4.103 \cdot 10^{-13}$
	$\rightarrow K^{-} + \mu^{+} + \nu_{\mu}$	3.19	
J/ψ	$\rightarrow \mu^{+} + \mu^{-}$	5.88	$\sim 10^{-20}$
etc.			

Note: The $\tau^{\pm}$ decay yields in addition a tau neutrino (${}^{(}\overline{\nu_{\tau}}{}^{)}$).

Due to the energy degradation of the hadron cascade in a shower as it penetrates into deeper regions of the atmosphere, the hadronic collisions become less energetic, and likewise the secondaries, emerging from the collisions, some of which are prospective parent particles of the muons. Therefore, muons resulting from later generations of interactions that occur at greater depth in the atmosphere are less energetic than those from the first few generations originating from great heights. The production of very high energy muons in the atmosphere is always associated with extensive air showers.

5.2. *Muon production via photopion production*

As an example for muon production via photopion production on nucleons, we take the case of a proton target,

$$\gamma + p \to n + \pi^+. \tag{10}$$

The threshold energy of the photon in the laboratory frame, $E_{ph,th,lab}$, needed for the above reaction to take place in a fixed target experiment is again obtained from the kinematic relation, which is in this case,

$$s = m_p^2 + 2m_p E_{ph,lab}. \tag{11}$$

This leads to the following expression for the threshold energy of a photon in the fixed target laboratory frame

$$E_{ph,th,lab} = m_\pi \left(1 + \frac{m_\pi}{2m_p}\right). \tag{12}$$

Inserting numbers for the masses we get for the threshold energy of this process, $E_{ph,th,lab} \simeq 145\,\mathrm{MeV}$.

The corresponding cross section is $\sigma_{\gamma,p} \simeq 6 \cdot 10^{-28}\,\mathrm{cm}^2$ near threshold and drops to $\simeq 10^{-28}\,\mathrm{cm}^2$ at a few GeV. From there it is still slowly declining with increasing energy, and just below 100 GeV it begins to rise again (for details see Olive *et al.*, 2014).[d,25]

---

[d]In two-body decays, the muon is completely polarized in the direction of motion of the muon in the rest frame of the meson (pion). The polarization of the muon is antiparallel (negative) to the direction of motion of the positively charged muon

5.3. Direct photoproduction of muon pairs

Photoproduction of a muon pair by a photon in the Coulomb field
of a proton or nucleus can be calculated analogously to electron pair
production on the basis of quantum-electrodynamics (QED).[26–29]
Symbolically, the reaction can be represented in its most simple form
without target excitation as

$$\gamma + p(\text{or } Z) \to p \ (\text{or } Z) + \mu^+ + \mu^-, \tag{13}$$

where γ, p, and Z represent the incident photon, the target proton,
or the target nucleus, respectively.

5.4. Muon production by neutrinos

Muon production in neutrino initiated interactions is very rare
because of the small cross sections, and of no concern in the context
of "Muography".

6. Muon Decay

The muon (μ^-) and its antiparticle (μ^+) are relatively long-lived
unstable particles with a mean life of $2.1970 \cdot 10^{-6}$ s at rest.
They decay through weak interaction, following almost exclusively
($\sim$100%) the decay scheme

$$\mu^- \to e^- + \nu_\mu + \overline{\nu_e}, \tag{14}$$

and

$$\mu^+ \to e^+ + \overline{\nu_\mu} + \nu_e. \tag{15}$$

7. Muon Interactions and Energy Losses

7.1. General comments

Muons propagating in matter (solids, liquids, or gases) are subject to
the following energy loss mechanisms: *ionization* (including *atomic*

(μ^+), and parallel (positive) for the μ^- in the case of π^--decay. The degree of
polarization of the muon in the laboratory frame depends on the velocity of the
parent meson, and on the direction of emission of the muon with respect to the
direction of motion of the meson.

excitation and *knock-on electrons*),[30–32] *Bremsstrahlung*,[26,33,34] *direct electron pair production*,[35–37] and *photonuclear interactions*.[27,38–47] In addition, higher-order effects from ionization[48] and direct muon pair production[49] make rare contributions to the energy loss.

The references listed above are those of the basic papers where the models and computational approaches to treat the particular process are discussed. Reviews of these subjects are found in Crispin and Fowler (1970),[50] Bugaev *et al.* (1970),[51] Kotov and Logunov (1970),[52] Bergamasco and Picchi (1971),[53] Vavilov *et al.* (1974),[54] Grupen (1976),[55] Lohmann *et al.* (1985),[56] and Bugaev *et al.* (1993 and 1994).[57,58] Direct muon pair production and other higher-order processes are usually disregarded unless specific aims are pursued.

The total energy loss formula for muons can be written as

$$-\frac{dE}{dx} = a_{ion}(E) + [b_{br}(E) + b_{pp}(E) + b_{ni}(E)]E. \qquad (16)$$

The term $a_{ion}(E)$ in Eq. (16) stands for energy losses due to *ionization, atomic excitation*, and *knock-on electrons*. It has a weak logarithmic energy dependence and is therefore sometimes regarded as quasi-constant for approximate range calculations and energy loss estimates in the mid-relativistic energy range.

The terms $b_{br}(E)$, $b_{pp}(E)$, and $b_{ni}(E)$ represent the energy losses resulting from *Bremsstrahlung, pair production*, and *photonuclear interactions*, respectively. Each of these mechanisms is energy-dependent and the mathematical expressions are relatively complex. Different approaches and approximations were chosen by different authors. In the following sections, we summarize the different processes based on the work of the references listed above and use in parts the notation of Kokoulin and Petrukhin (1970).[49]

In Table 3, we give as an example the values of the different terms listed in Eq. (16) for 6 energies of muons propagating in iron. The values are taken from the tables of Lohmann *et al.* (1985).[56]

In the past, several authors have given approximate expressions for $a_{ion}(E)$, $b_{br}(E)$, $b_{pp}(E)$, and $b_{ni}(E)$ for quick computation of the average energy loss of muons for everyday use.[42,59] However, the deviations with respect to modern Monte Carlo simulations, e.g., Chirkin and Rhode (2016),[60] are large and we recommend to use the

Table 3.   Energy loss, dE/dx, of muons in iron [GeV g^{-1} cm^2] (after Ref. 56).

E [GeV]	a_{ion}	b_{br}	b_{pp}	b_{ni}	Total
1	$1.56 \cdot 10^{-3}$	$5.84 \cdot 10^{-7}$	$1.77 \cdot 10^{-7}$	$4.14 \cdot 10^{-7}$	$1.56 \cdot 10^{-3}$
10	$1.93 \cdot 10^{-3}$	$1.40 \cdot 10^{-5}$	$1.49 \cdot 10^{-5}$	$4.23 \cdot 10^{-6}$	$1.96 \cdot 10^{-3}$
100	$2.16 \cdot 10^{-3}$	$2.24 \cdot 10^{-4}$	$3.17 \cdot 10^{-4}$	$3.85 \cdot 10^{-5}$	$2.74 \cdot 10^{-3}$
300	$2.25 \cdot 10^{-3}$	$7.72 \cdot 10^{-4}$	$1.13 \cdot 10^{-3}$	$1.14 \cdot 10^{-4}$	$4.27 \cdot 10^{-3}$
1000	$2.45 \cdot 10^{-3}$	$2.87 \cdot 10^{-3}$	$4.19 \cdot 10^{-3}$	$3.88 \cdot 10^{-4}$	$9.78 \cdot 10^{-3}$
10000	$2.50 \cdot 10^{-3}$	$3.17 \cdot 10^{-2}$	$4.52 \cdot 10^{-2}$	$4.33 \cdot 10^{-3}$	$8.38 \cdot 10^{-2}$

tabulated energy losses of Lohmann et al. (1985)[56] instead for quick reference.

7.2.   Ionization losses of muons

As an example, for a high-energy muon traversing the atmosphere in vertical direction ($\simeq$1030 g/cm^2) the ionization losses amount to about 2.2 GeV (Rossi, 1941 and 1952; Barnett et al., 1996).[30, 61, 62] For such particles, the rate of energy loss by ionization varies *logarithmically* with energy (for details see Fig. 4).

The ionization cross-section of a muon incident on an atom of atomic number Z can be written as[30]

$$\left(\frac{d\sigma}{dv}\right)_{ion} = Z\left(\frac{2\pi r_e^2}{\beta^2 v^2}\right)\left(\frac{m_e}{E_\mu}\right)\left[1 - \beta^2\left(\frac{v}{v_{max}}\right) + \frac{v^2}{2}\right], \qquad (17)$$

where Z is the atomic number of the target medium, r_e, the classical radius of the electron, m_e, its rest mass, v, the fractional energy of the muon transferred to the electron, E_μ, the energy of the incident muon, and $\beta = v/c$ is its velocity in terms of the velocity of light, c. The units for mass, energy, and momentum are GeV and $c = 1$. This equation is valid, provided that the momentum transfer to the atomic electron is large enough that it can be considered to be free of all bound state and screening effects.

The maximum fractional energy transfer is

$$v_{max} = \frac{\beta^2}{1 + \left(\frac{m_\mu^2 + m_e^2}{2 m_e E_\mu}\right)}, \qquad (18)$$

where m_μ is the muon mass.

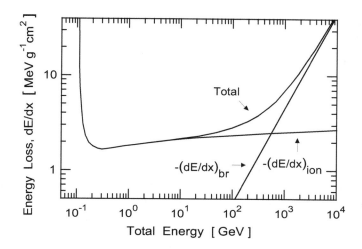

Figure 4. Energy loss of muons versus total energy in standard rock ($Z = 11$), due to ionization and Bremsstrahlung.

The average incremental energy loss by ionization per unit path length in a medium is given by the Bethe–Bloch formula[30,63] which can be written as

$$
-\left(\frac{dE}{dx}\right)_{ion} = 2\pi r_e^2 \left(\frac{N_A Z}{A}\right) \left(\frac{m_e}{\beta^2}\right)
$$
$$
\cdot \left[\ln\left(\frac{2m_e E_\mu \beta^2 v_{max}}{(1-\beta^2)I^2(Z)}\right) - 2\beta^2 + \left(\frac{v_{max}^2}{4}\right) - \delta \right]. \tag{19}
$$

Here, x is the path length expressed in [$\mathrm{g\,cm^{-2}}$], i.e., the mass per unit area or the *column density* of the target, r_e, the classical electron radius (2.817^{-13} cm), N_A, Avogadro's number ($6.023 \cdot 10^{23}$), Z and A are the atomic number and the mass number of the target, m_e, the mass of the electron, $\beta = p/E_\mu$, where p is the muon momentum, $\gamma = E_\mu/m_\mu$ the Lorentz factor of the muon, $I(Z)$, the mean ionization potential of the target ($I(Z) \sim 16Z^{0.9}$ [eV] for $Z \geq 1$), and δ is the density effect which approaches $2\ln(\gamma)$.[31,64] Note that the high energy tail of the distribution yields high energy knock-on electrons that initiate electromagnetic (EM) cascades.

Equation (19) can be rewritten in a somewhat more explicit form as

$$-\left(\frac{dE}{dx}\right)_{ion} = 2\pi N_A \alpha^2 \lambda_e^2 \left(\frac{Z}{A}\right)\left(\frac{m_e}{\beta^2}\right)$$
$$\cdot \left[\ln\left(\frac{2m_e\beta^2\gamma^2 E'_{max}}{I^2(Z)}\right) - 2\beta^2 + \left(\frac{E'^2_{max}}{4E_\mu^2}\right) - \delta\right],$$
(20)

where α is the fine-structure constant (1/137), λ_e, the Compton wavelength of the electron (3.86 · 10^{-11} cm), and E'_{max} is the maximum energy transferable to the electron,

$$E'_{max} = 2m_e\left(\frac{p^2}{m_e^2 + m_\mu^2 + 2m_e\sqrt{(p^2 + m_\mu^2)}}\right).$$
(21)

For more accurate calculations, the following expressions should be used for δ:

$$\delta(X) = 4.6052X + a(X_1 - X)^m + C \text{ for } X_0 < X < X_1 \text{ and} \quad (22)$$

$$\delta(X) = 4.6052X + C \text{ for } X > X_1. \quad (23)$$

Here, $X = \lg(\beta\gamma)$. The values of X_0, X_1, a, m, C, and $I(Z)$ are given in the tables of Sternheimer (1956)[31] and Sternheimer *et al.* (1984).[64]

7.3. *Muon Bremsstrahlung*

This process, symbolically written in the following expression (24), had first been calculated by Bethe and Heitler (1934)[26] and is very similar to electron bremsstrahlung.

$$\mu + p \text{ (or } Z) \to \mu' + p \text{ (or } Z) + \gamma. \quad (24)$$

Since the emission of photons by muons takes place at much smaller impact parameters from the nucleus than in the case of electrons, the screening of the nuclear Coulomb field by the outer electrons can be neglected to a greater extent than for electrons. At distances less than the nuclear radius, the field cannot be regarded as a point charge and the spin of the particles becomes significant. According to

Christy and Kusaka (1941a and 1941b),[65,66] the differential radiation probability for muons (spin 1/2 particles) per gram per square centimeter [g^{-1}cm^2], $P(E_\mu, v)_{\mu br}$, where v is the fractional energy of the emitted photon, is given by the following expression (see also Ref. 61):

$$P(E_\mu, v)_{\mu br}\, dv = \alpha \frac{N_A}{A} Z^2 r_e^2 \left(\frac{m_e}{m_\mu}\right)^2 \frac{16}{3}\left(\frac{3v}{4} + \frac{(1-v)}{v}\right) dv$$

$$\cdot \left[\ln\left(\frac{12}{5}\frac{(1-v)}{v}\frac{E_\mu}{m_\mu Z^{\frac{1}{3}}}\right) - \frac{1}{2}\right]. \tag{25}$$

Here, E_μ is the total energy of the incident muon, m_μ, its rest mass, and m_e, the rest mass of the electron. The remaining symbols are as defined earlier. The mass ratio term suppresses the probability for radiation by the factor $(m_e/m_\mu)^2$, which amounts to more than four orders of magnitude. From Eq. 25, the cross section for Bremsstrahlung of a muon in the Coulomb field of a proton or nucleus can be computed. Note that Eq. (25) is valid only for $E \gg m_\mu$.

Several authors have chosen different approaches and approximations to derive the cross section of muon Bremsstrahlung. In the following, we present the expression obtained by Petrukhin and Shestakov (1968),[33] which is widely used today.[56,67]

$$\left(\frac{d\sigma}{dv}\right)_{br} = \alpha \left(2Z r_e \frac{m_e}{m_\mu}\right)^2 \frac{1}{v}\left[1 + (1-v)^2 - 2\left(\frac{1-v}{3}\right)\right]\phi(q_{min}),$$

$$\tag{26}$$

where

$$\phi(q_{min}) = \ln\left[f_n\left(\frac{m_\mu}{m_e}\right)\left(\frac{RZ^{-1/3}}{1 + (q_{min}/m_e)\sqrt{e}RZ^{-1/3}}\right)\right], \tag{27}$$

and

$$f_n = \left(\frac{2}{3}\right)Z^{-1/3}, \tag{28}$$

is the nuclear form factor correction, $e = 2.7182...$, and $R = 189$. Furthermore,

$$q_{min} = \frac{m_\mu^2 v}{2E_\mu(1 - v)} \tag{29}$$

is the minimum momentum transfer to the proton or nucleus. The range of the fractional energy transfer, v, is

$$0 < v \le \left(1 - \frac{m_\mu \sqrt{e}}{2f_n E_\mu}\right). \tag{30}$$

The Petrukhin–Shestakov cross-section (Eq. (26)) is an analytic approximation to the Bethe–Heitler cross-section with arbitrary screening. Sakumoto et al. (1992)[67] find in their work that the function f_n of Petrukhin and Shestakov (1968)[33] underestimates the numerical calculation by about 10% and propose to use in its place the expression $f_n = \exp(-0.128R_{0.5})$ with $R_{0.5} = (1.18A^{1/3} - 0.48)$ [fm].

As an example, we show in Fig. 4 the energy dependence of the two major processes that are responsible for the energy loss of muons over the range $0.1 \le E_\mu \le 10^4$ GeV in standard rock ($\langle Z \rangle = 11$, $\langle A \rangle = 22$, $Z/A = 0.5$, $Z^2/A = 5.5$, $\rho = 2.650$ g cm^{-3}), and the sum of both (for details see Ref. 68).

In Fig. 5, we show the same plot for the energy loss of muons in air with the contributions of the different energy loss mechanisms indicated[69, 70]

The photon resulting from Bremsstrahlung of a high-energy muon in the Coulomb field of a nucleus can initiate an EM shower that accompanies the muon trajectory over a certain track length, depending on its energy, and the medium in which it propagates. Such EM-cascades as well as photonuclear reaction products of high-energy muons can regenerate horizontal air showers at great atmospheric depth. They also contaminate a pure muon beam and are of particular relevance for highly shielded deep underground experiments that require low background, or measurements of high-energy muons under thick absorbers, or in calorimeters in a shower core. In the latter case, so-called *punch-throughs* pose an additional problem.

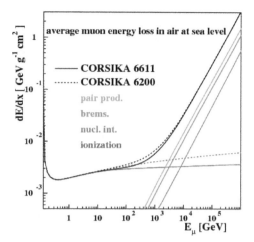

Figure 5. Average energy loss of muons versus energy in air ($Z = 7.3$), due to different interactions (after Ref. 69).

The mean energy loss is obtained by evaluating the expression

$$\frac{dE}{dx} = E_\mu \left(\frac{N_A}{A}\right) \int_{v_{min}}^{v_{max}} \left(\frac{d\sigma}{dv}\right) dv, \tag{31}$$

where the range of v is given by Eq. (30), i.e., $v_{min} = 0$ and for v_{max} we can write

$$v_{max} = 1 - \left(\frac{3}{4}\right) \sqrt{e} \left(\frac{m_\mu}{E}\right) Z^{1/3}. \tag{32}$$

To account for Bremsstrahlung on atomic electrons, one usually replaces in Eq. (26) Z^2 by $Z(Z+1)$.[34,35]

7.4. *Direct electron pair production by muons*

The differential cross section for the direct production of electron pairs by muons in the Coulomb field of a proton or nucleus, shown symbolically as follows:

$$\mu^\pm + p \ (\text{or} \ Z) \ \rightarrow \ \mu^{\pm'} + p \ (\text{or} \ Z) + e^+ + e^-, \tag{33}$$

can be calculated under the assumption that all particles are relativistic, and that the Thomas–Fermi model can be used to compute the effect of screening of nuclei by atomic electrons. Kelner

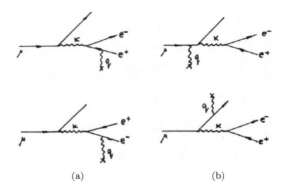

(a)   (b)

Figure 6. Diagrams (a) and (b) represent the pair production process in the lowest order of perturbation theory as used by Kelner (1967)[71] and Kelner and Kotov (1968b).[35]

and Kotov (1968a and 1968b)[34, 35] derived expressions for the two extreme cases, *no screening* and *complete screening*.

Using the Thomas–Fermi model and following the steps of Kelner (1967)[71] in conjunction with the interpolation formulas of Petrukhin and Shestakov (1966),[72] one obtains the following expression for the differential cross section for diagram (a) of Fig. 6:

$$d\sigma_a = \left(\frac{16}{\pi}\right) (Z\alpha r_e)^2 F_a(E_\mu, v) \frac{dv}{v}. \tag{34}$$

For $v \sim 1$, the contribution of diagram (b) is of the same order of magnitude as diagram (a) and one obtains an equation for $d\sigma_b$ analogous to Eq. (34) for (b) with $F_b(E_\mu, v)$. The functions $F_a(E_\mu, v)$ and $F_b(E_\mu, v)$ are evaluated by integration over x and t, where $x_\pm = \epsilon_\pm/\omega$, and the interference between (a) and (b) becomes zero.

This calculation is valid provided that the energies of the particles in the initial and final states are much larger than their masses, i.e., when $E_\mu \gg m_\mu$, $x_\pm \gg 1/E_\mu v$, where v and $(1 - v)$ are the energy fractions of the electrons, $v \gg 2/E_\mu$, and $(1 - v) \gg m_\mu/E_\mu$.

Finally, the complete differential cross-section for direct electron pair production by muons can be written as

$$d\sigma = \left(\frac{16}{\pi}\right) (Z\alpha r_e)^2 F(E_\mu, v) \frac{dv}{v}, \tag{35}$$

where $F(E_\mu, v) = F_a + F_b$. The function $F(E_\mu, v)$ is plotted in Fig. 7.

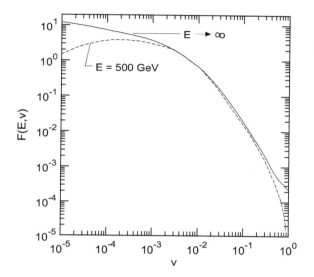

Figure 7. Example of the function $F(E_\mu, v)$ for standard rock ($Z = 11$) plotted as a function of v for muon energies $E_\mu = 500\,\text{GeV}$ and $E_\mu \to \infty$. For the detailed calculation and tabulated values, see Kelner and Kotov (1968b).[35]

From Eq. (35), the formula for the energy loss of muons for $e^+ + e^-$ pair production can now be obtained as

$$-\frac{1}{E_\mu}\frac{dE_\mu}{dx} = \left(\frac{16}{\pi}\right)(Z\alpha r_e)^2 n \cdot \chi(E_\mu),\qquad (36)$$

where

$$\chi(E_\mu) = \int_{v_{min}}^{v_{max}} F(E_\mu, v)\,dv,\qquad (37)$$

and n is the number of atoms per unit volume. Kelner and Kotov have computed the values of $\chi(E_\mu)$ for a wide range of energies and present tabulated data in their paper.

For the two extreme cases, *negligible screening* and *complete screening*, the following simple equations result:

$$\left(-\frac{1}{E_\mu}\frac{dE_\mu}{dx}\right)_{pair} = n\left(\frac{19\pi}{9m_\mu}\right)(Z\alpha r_e)^2 \cdot \left[0.965\ln\left(\frac{E_\mu}{4m_\mu}\right) - 1.771\right],$$

$$(38)$$

and

$$\left(-\frac{1}{E_\mu}\frac{dE_\mu}{dx}\right)_{pair} = n\left(\frac{19\pi}{9m_\mu}\right)(Z\alpha r_e)^2 \cdot \left[0.965\ln(189Z^{1/3}) + 0.605\right].$$ (39)

For muons complete screening applies if their energy is $\geq 10^{13}$ eV. In this case and for $Z = 11$ (standard rock), one obtains the following simple approximate expression for the energy loss:

$$\left(-\frac{1}{E_\mu}\frac{dE_\mu}{dx}\right)_{pair} = 2.40 \cdot 10^{-6} \text{ [g}^{-1}\text{ cm}^2\text{]}.$$ (40)

Figure 8 shows a plot of the energy loss due to electron pair production by muons in standard rock.

We should point out that in a different approach, based on the work of Kelner and Kotov (1968a and 1968b),[34,35] Kokoulin and Petrukhin (1970)[49] derived another frequently used alternative approximation for the differential cross-section and the energy loss.

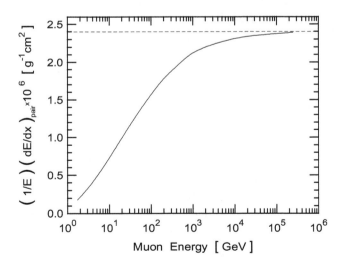

Figure 8. Energy loss of muons through direct electron pair production in standard rock ($Z = 11$) as a function of muon energy.[35]

7.5. *Direct muon pair production by muons, muon trident events*

Direct production of muon pairs by high energy muons in the Coulomb field of a proton or nucleus in reactions such as the following:

$$\mu^{\pm} + p \text{ (or } Z) \;\rightarrow\; \mu^{\pm'} + p \text{ (or } Z) + \mu^{+} + \mu^{-}, \tag{41}$$

had been observed in cosmic ray and accelerator experiments.[73–75] The energy loss mechanism of this process had been studied by several authors and had been accounted for in muon range calculations.[49,56,67]

The cross section obtained by Russell *et al.* (1971)[75] in their accelerator experiment with a positive and negative muon beam of 10.5 GeV on a Pb target is

$$\sigma(\mu^{\pm}, Pb) = 51.7 \pm 7 \cdot 10^{-33} \text{ [cm}^2/\text{Nucleus]}, \tag{42}$$

in agreement with QED.

7.6. *Photonuclear interactions of muons*

Nuclear interactions of muons (or electrons) with nucleons or nuclei, i.e., inelastic reactions where hadrons are being produced, are so-called *photonuclear interactions* where virtual photons are involved. An example of such a reaction is as follows:

$$\mu^{\pm} + p \text{ (or } A) \;\rightarrow\; \mu^{\pm'} + p \text{ (or } A) + X, \tag{43}$$

where X stands for hadrons (pions, etc.).

These processes are theoretically not so well understood as the processes where real photons are involved. Photonuclear processes initiated by muons play an important role in highly shielded deep underground experiments where they cause problematic backgrounds, and occasionally in thick calorimeters.

A summary of the results of early measurements of photonuclear interaction cross-sections of muons obtained from accelerator and cosmic ray experiments on nuclear targets, expressed per nucleon, is given in Fig. 9.

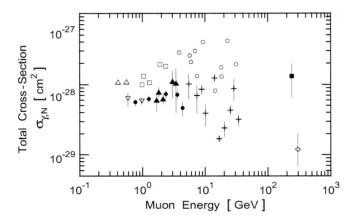

Figure 9. Data summary of early measurements of the energy dependence of the photonuclear interaction cross-section of muons (Refs. 46 and 81 and references listed therein). The plot includes accelerator gamma ray data by Chasan et al. (1960)[82] and Crouch et al. (1964) (•);[83] accelerator muon data by Kirk et al. (1965)[84] interpreted using the Weizsäcker–Williams (WW)-theory ($\triangle$) and the Kessler–Kessler (KK)-theory for the same data ($\triangledown$). The cosmic ray muon data include the analysis of Fowler and Wolfendale (1958)[85] using the WW-theory ($\square$) and the KK-theory ($\blacktriangle$). The symbols $\circ$ and + are data from different experiments evaluated according to the WW and KK theories, respectively, and likewise for points $\blacksquare$ and $\diamond$, which are from the work of Borog et al. (1968)[81] using cosmic ray muons.

The theoretical treatment of these interactions are based on the exchange of a virtual photon.[38–41, 76–79] Using Lorentz and gauge invariance only, the cross-section for hadron production by muons (or electrons) can be expressed by two Lorentz-invariant functions where the character of the strong interaction is confined.[e, 42, 59, 80]

Following Kobayakawa (1967)[42] and using his notation (not the Feynman notation convention), then, when an incident muon (electron) with 4-momentum $Q_1(\vec{p}_1, E_1)$ collides with a nucleon at rest $(Q(0, M))$ and has a 4-momentum $Q_2(\vec{p}_2, E_2)$ in the final state, emitting a virtual photon with 4-momentum $q(\vec{q}, \epsilon)$, the *differential* cross-section, expressed as a function of the transferred energy ϵ and

---

[e]In the early work of Higashi et al. (1965),[80] the criterion for hadron production was the production of a hadronic cascade of energy $\geq E$ by a muon.

the square of 4-momentum transfer,

$$q^2 = |\vec{q}|^2 - \epsilon^2 > 0, \text{ is} \tag{44}$$

$$\frac{d^2\sigma}{dq^2\,d\epsilon} = \frac{\alpha}{8\pi^2} \frac{1}{(E_\mu^2 - m_\mu^2)} \frac{1}{q^4}$$

$$\cdot \left[L_N \left((E_\mu^2 + (E_\mu - \epsilon)^2)q^2 - 2m_\mu^2\epsilon^2 - \frac{q^4}{2} \right) \right. \tag{45}$$

$$\left. + L_N'(2m_\mu^2 - q^2)q^2 \right]$$

where α is the fine-structure constant and m_μ the muon mass. L_N and L_N' are functions of q^2, $p^2(= -M^2)$, and $qp(= -M\epsilon)$, and the contribution of the strong interaction is confined to these functions. With the assumptions of the forms of the L and L' functions, a practical expression of the differential cross-section can be obtained. For the detailed treatment, the reader is referred to the paper of Kobayakawa (1967)[42] and references listed therein.

More recently, Bezrukov and Bugaev (1981)[47] derived a more elaborate expression for the photonuclear cross-section, which is now used frequently.

8. Summary of Muon Reaction Probabilities and Energy Losses

Figure 10 shows the energy loss probability of muons for different processes in standard rock obtained from an early calculation of Adair (1977).[59]

A very comprehensive set of muon energy loss data for many different elements and compounds, including standard rock and water, are given in tabulated form in the report of Lohmann *et al.* (1985),[56] covering the energy range from 1 to 10^4 GeV. Bhattacharyya (1986)[86] presents in his paper plots of the energy loss of muons in the range from 50 GeV to 1,000 GeV in sea water that are compared with experimental data.

The energy loss of cosmic ray muons in iron covering the energy range from 40 GeV to 1,200 GeV had been measured by

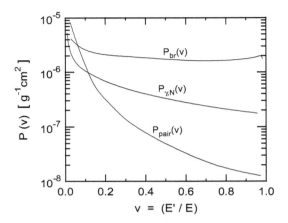

Figure 10. Probability of a muon of energy E_μ losing a fraction of its energy $v = E'_\mu/E_\mu$ per gram of *standard rock* traversed through Bremsstrahlung, $P_{br}(v)$, photonuclear reaction, $P_{\gamma,N}(v)$, and direct electron pair production, $P_{pair}(v)$ (after Ref. 59). The curves apply to 2 TeV muons but are valid over a wide range of energies.

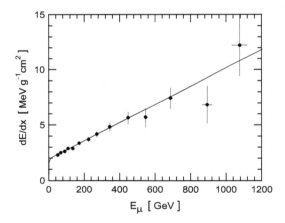

Figure 11. The dE_μ/dx-plot for muons in iron as a function of energy. The solid curve is the prediction from a calculation (after Ref. 67).

Sakumoto et al. (1992)[67] in a precision experiment. Their work had been complemented by detailed calculations using the equations given in the previous subsections. The results are shown in Fig. 11.

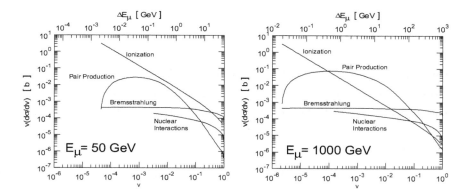

Figure 12. Ionization, Bremsstrahlung, direct electron pair production and photonuclear cross sections (in units of barns) of 50 GeV and 1,000 GeV muons incident on an iron atom. The scale on top of the figure gives the energy loss, ΔE_μ which corresponds to the fractional energy loss, v. Note that the ordinate is the logarithmic derivative $d\sigma/d(\ln v) = d\sigma/d(\ln \Delta E_\mu)$ (after Ref. 67).

Table 4. Photonuclear reaction cross sections of muons from cosmic ray experiments.

Experimental	E_μ [GeV]	References
$\sigma(\mu^\pm, Pb)^* = (2.58 \pm 0.36) \cdot 10^{-31}$ [cm²]	≥ 100	Higashi *et al.* (1965)[80]
$\sigma(\mu^\pm, Fe)^* = (3.62 \pm 0.76) \cdot 10^{-31}$ [cm²]	≥ 100	

Experimental [g⁻¹ cm² GeV⁻¹] in Standard Rock (SR)	E_μ [GeV]	References
$\sigma(\mu^\pm, SR) = (1.9 \pm 0.1) \cdot 10^{-10}$	≥ 190	Enikeev *et al.* (1983)[96]
$\sigma(\mu^\pm, SR) = (7.6 \pm 0.6) \cdot 10^{-11}$	≥ 270	
$\sigma(\mu^\pm, SR) = (8.1 \pm 1.4) \cdot 10^{-12}$	≥ 550	
$\sigma(\mu^\pm, SR) = (2.2 \pm 0.6) \cdot 10^{-12}$	≥ 780	
$\sigma(\mu^\pm, SR) = (5.3 \pm 2.3) \cdot 10^{-13}$	≥ 1320	
$\sigma(\mu^\pm, SR) = (1.5 \pm 1.1) \cdot 10^{-13}$	≥ 1860	

Note: *Cross sections refer to penetrating shower-producing events.

Additional results from the theoretical work of Sakumoto *et al.* (1992)[67] that are of interest for many applications are reproduced in Fig. 12. In Table 4, we list some cross sections obtained from cosmic ray experiments.

9. Zenith Angle Dependence of the Atmospheric Column Density

In a standard isothermal exponential atmosphere that is characterized by a constant scale height $h_s = (kT/Mg)$ [cm], where k is Boltzman's constant, T [K], the temperature in Kelvin, M [g/mol], the molecular weight, and g [cm^{-1}s^{-2}], the gravitational acceleration, the vertical column density X [g/cm^2] of air overlaying a point P at altitude h [cm] is given by the common *barometer formula*

$$X(h) = X(h = 0)e^{-(h/h_s)} \ [\text{g/cm}^2]. \tag{46}$$

9.1. *Flat earth approximation*

For an inclined trajectory subtending a zenith angle $\theta \leq 60°$, the column density $X(h, \theta)$ measured from infinity to a given point P at altitude h in the atmosphere can be calculated neglecting the Earth's curvature, as shown in Fig. 13. In this case, the column density increases with respect to the vertical column density proportional to the secant of the zenith angle, θ. Thus,

$$X(h, \theta \leq 60°) = X(h, \theta = 0°) \sec(\theta) \ [\text{g/cm}^2]. \tag{47}$$

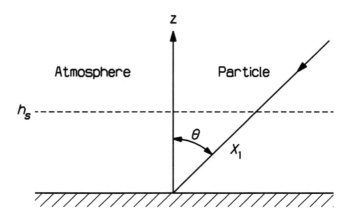

Figure 13. Atmospheric thickness or column density, X_1, encountered by a cosmic ray incident at a zenith angle θ to reach point P under the assumption that the Earth is flat. h_s represents the atmospheric scale height. (This approximation can be used for zenith angles $\theta \leq 60°$ to $\leq 75°$, depending on accuracy required.)

For less accurate calculations, this expression may even be used for zenith angles $\theta \leq 75°$.

9.2. *Curved earth atmosphere*

For larger zenith angles, the *curvature of the Earth* cannot be ignored as is evident from Fig. 14. The correct derivation of the formula to compute the true column density for inclined trajectories leads to the *Chapman function*.[87] The Chapman function gives the ratio of the total amount of matter along an oblique trajectory subtending a zenith angle θ versus the amount of total matter in the vertical ($\theta = 0$) for a given point P in the atmosphere. This function has the following form:

$$Ch(x, \theta) = x \sin(\theta) \int_0^\theta \frac{\exp(x - x \sin(\theta)/\sin(\phi))}{\sin^2(\phi)} \, d\phi, \qquad (48)$$

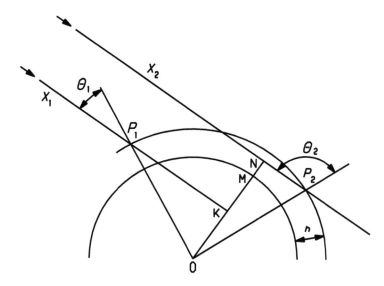

Figure 14. Atmospheric column density X_1 in curved atmosphere encountered by a cosmic ray incident under zenith angle $\theta_1 \leq \pi/2$ to reach point P_1 at altitude h. Also shown is the situation for point P_2 at $\theta > \pi/2$ and column density X_2, a situation that may arise when h is large.

where

$$x = \frac{R_E + h}{h_s}. \tag{49}$$

R_E is the radius of the Earth, h, the altitude of point P in the atmosphere, and h_s is the appropriate scale height of the atmosphere.

Various authors have derived approximations for this expression.[88,89] Swider and Gardner propose for zenith angles $\theta \leq (\pi/2)$ the equation

$$Ch\left(x, \theta \leq \frac{\pi}{2}\right) = \left(\frac{\pi x}{2}\right)^{1/2} \left(1 - erf\left[x^{1/2} \cos\frac{\theta}{2}\right]\right) \exp\left(x \cos^2\frac{\theta}{2}\right). \tag{50}$$

For $\theta = \pi/2$, i.e., for horizontal direction,

$$Ch\left(x, \frac{\pi}{2}\right) = (\pi x/2)^{1/2}, \tag{51}$$

which is about equal to 40.

For zenith angles in excess of $\pi/2$, a situation that may arise in satellite experiments (cf Fig. 14), the same authors propose the approximation

$$Ch\left(x, \theta \geq \frac{\pi}{2}\right) = \left(\frac{\pi x}{2}\sin(\theta)\right)^{1/2}\left(1 + erf\right.$$

$$\times \left[-\cot(\theta)\left(\frac{x\sin(\theta)}{2}\right)^{1/2}\right]\right)$$

$$\times \left(1 + \frac{3}{8x\sin(\theta)}\right). \tag{52}$$

The zenith angle dependence of the atmospheric thickness or column density at sea level is illustrated in Fig. 15.

For further details concerning atmospheric column densities and attenuation for zenith angles $\theta \geq \pi/2$, such as may be relevant at great altitude in conjunction with satellites, the reader is referred to the articles by Swider (1964)[90] and Brasseur and Solomon (1986).[91] The accuracy of certain approximations for the Chapman function is discussed by Swider and Gardner (1967).[89]

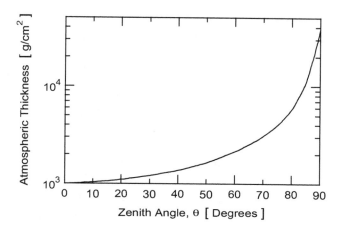

Figure 15. Relation between zenith angle and atmospheric thickness or column density at sea level for the "curved" Earth.

10. Muon Data at Sea Level

10.1. *General comments*

It is important to realize that for any accurate calculation of the muon flux after penetrating a (homogeneous) target (absorber), or for the interpretation of the measured muon flux after the penetration of an unknown target, e.g., to deduce the existence of voids within it, the incident muon flux, i.e., the flux before the target, must be determined carefully. This is particularly important for relatively thin targets where muons of lower energy are being used as their flux is more location-dependent than the high energy flux.

There exists a vast amount of data on muons from all altitudes in the literature. However, the bulk of the data was recorded in the lower regions of the atmosphere, the majority near or at sea level. The measurements were usually made with solid iron magnetic spectrometers.[92–94] For a very comprehensive compilation of muon data in the atmosphere, at sea level, and underground, the reader is referred to the book *Cosmic Rays at Earth*.[95]

The muon component in the atmosphere had been investigated most thoroughly during the past several decades, practically since the discovery of the muon and in fact even before, under the

name "penetrating particles" or "penetrating component". A large
number of experiments carrying out countless measurements of ever
higher quality and accuracy at many different locations around
the globe have contributed muon data of all kinds (differential
and integral fluxes, momentum and energy spectra, charge ratios,
zenith and azimuthal distributions, etc.). The great wealth of data
is well-documented and many valuable reviews and data archives
are now available to users. Note that muon data and spectra are
frequently specified in units of momentum (GeV/c) since magnetic
spectrometers yield directly the momentum of the particles and not
the energy. The latter must be calculated from the former.

Since some time a certain saturation of high quality muon
data has been reached and the activities in this field of research
have diminished significantly. This is the reason why there are few
references of recent date covering the muon momentum range in the
lower momentum domain (≤ 1000 GeV/c). Most of the current work
on muons focuses on high and ultrahigh energy (UHE) muons in
the TeV and PeV energy range, i.e., in an energy region where the
event rate is very low and only giant detectors can make reasonable
contributions to new spectral regions. Today the giant IceCube
detector located at the South Pole is the main contributor of data
on UHE muons.[70]

It is important to note that up to a muon momentum of about
5 GeV/c, the muon flux varies from geographic location to location
because of the location-dependent geomagnetic field that imposes
a geographic (geomagnetic) latitude dependent cutoff momentum,
P_c, on the primary cosmic radiation. The location dependence of
the muon flux increases with decreasing muon momentum. This is
well documented by Table 5, which shows a compilation of absolute
integral intensities of muons above a given threshold momentum
at or near sea level, collected at different locations on the globe. The
large spread of intensities at low momenta is the principal reason why
most of the muon measurements terminate at momenta of about 0.2
or 0.3 GeV/c; such low energy data are of no general interest.

Table 5. Vertical absolute integral intensities of muons at or near sea level (ordered by latitude).

Authors	Geomagn. Lat. [°N]	P_c^a [GV]	Altitude [m]	Momentum [GeV/c]	Intensity $\times 10^3 [\mathrm{cm}^{-2}\, \mathrm{s}^{-1}\mathrm{sr}^{-1}]$
Allkofer et al. (1968)[97]	9	14.1	s.l.	≥0.32	7.25 ± 0.1
Chandrasekharan et al. (1950)[98]	9	—	555	≥0.27	7.6
Sinha and Basu (1959)	12	16.5	30	≥0.27	7.3 ± 0.2
De et al. (1972)	12	16.5	30	≥0.954	6.86 ± 0.03
Karmakar et al. (1973)	16	15.0	122	≥0.353	8.99 ± 0.05
				≥1.0	6.85 ± 0.04
Gokhale (1953)	19	—	—	≥0.32	7.3 ± 0.1
Gokhale and Balasubrahmanyam	19	—	124	≥0.27	7.55 ± 0.1
Fukui et al. (1957)	24	12.6	s. l.	≥0.34	7.35 ± 0.20
				≥0.54	6.87 ± 0.25
Kitamura and Minorikawa (1953)	25	12.6	—	≥0.34	7.2 ± 0.1
Baschiera et al. (1979)	42	4.5	238	≥0.457	8.75 ± 0.33
				≥0.918	7.27 ± 0.26
Wentz et al. (1995)	44.5	3.4	116	≥0.6	8.54 ± 0.34
Rossi (1948)[99]	≥50	~1.8	s.l.	≥0.32	8.3
Pomerantz (1949)	52	2.0	89	≥0.31	8.2 ± 0.1
Allkofer et al. (1970a, b, and 1971d erratum)[100–102]	53	2.4	s. l.	≥0.985	7.49 ± 0.30
				≥1.239	6.76 ± 0.27
Allkofer et al. (1971b, c),[103, 104]				≥0.4²)	9.18
Allkofer & Clausen (1970)[105]				≥1.0	7.22
Kraushaar (1949)	53	1.6	259	≥0.28	8.87 ± 0.05
Greisen (1942)	54	1.5	259	≥0.33	8.2 ± 0.1
			s. l.	≥0.33	8.3 ± 0.1^b)

(*Continued*)

Table 5. (*Continued*)

Authors	Geomagn. Lat. [°N]	P_c^a [GV]	Altitude [m]	Momentum [GeV/c]	Intensity $\times 10^3$ [cm^{-2} s^{-1}sr^{-1}]
Allkofer (1965)[106]	55	2.2	s. l.	$\geq$0.320	8.5 ± 0.2
Allkofer (1965)[106]				$\geq$0.320	8.4 ± 0.1
Crookes & Rastin (1972)	53	2.5	40	$\geq$0.35	9.13 ± 0.12
Crookes and Rastin (1971b, 1973)				>7.3	1.40 ± 0.02
				>8.5	1.19 ± 0.06
Barbouti and Rastin (1983)[108]	53	2.5	40	$\geq$0.438	$8.868 \pm 1.3\%$
				$\geq$0.815	$7.661 \pm 1.1\%$
				$\geq$1.728	$5.563 \pm 1.1\%$
				$\geq$2.681	$4.152 \pm 1.2\%$
				$\geq$3.639	$3.20 \pm 1.1\%$
Hayman et al. (1962)	57.5	1.8	s.l.	$\geq$0.320	$\geq 7.6 \pm 0.06$
Ashton et al. (1972)	57.5	2.1	s. l.	$\geq$0.88	8.22 ± 0.4
				$\geq$1.0	7.58 ± 0.4^c
Ayre et al. (1971a,b, and 1973a)	57.5	2.1	s. l.	>3.48	2.86 ± 0.04
				>7.12	1.31 ± 0.02

Notes: [a]Cutoff rigidities, P_c, are listed as given in references. They may vary with time and model employed.
[b]Results after corrections made by Rossi (1948).[99]
[c]Obtained by combining with the measurements of Allkofer et al. (1970a)[100] and Allkofer and Clausen (1970)[105] around 1 GeV/c. For references see Grieder (2001).[95]

In addition, the Earth's magnetic field gives rise to an *azimuthal dependence* and an *east–west asymmetry* of the muon intensity, because the primary cosmic ray particles are predominantly positively charged. It is mostly the low energy component that is affected to the extent that the path lengths are different for charged particles coming in from the east or west. As a result, the absorption and decay probabilities of unstable particles are influenced and the low energy muon charge ratio is affected. Besides the latitude and the

Table 6. Low momentum vertical differential muon intensities at sea level.

Authors	Cutoff P_c [GV]	Momentum [GeV/c]	Intensity $\times 10^3$ [cm^{-2}s^{-1}sr^{-1} (GeV/c)$^{-1}$]
Rossi (1948)[99]	~1.8	1.0	2.45
Ng *et al.* (1974b)	2.1	0.85 ± 0.03	4.09 ± 0.21
		1.16 ± 0.04	3.29 ± 0.19
Allkofer *et al.* (1970d)	2.4	1.32	2.57 ± 0.21
Allkofer *et al.* (1970a,e, and 1971d erratum)[100, 102]		1.0	$3.21 \pm 5\%$
		1.112	2.90 ± 0.2
Allkofer and Clausen (1970)[105]		1.24	2.73 ± 0.23
Allkofer and Jockisch (1973)		1.0	$3.09 \pm 8\%$*
Baschiera *et al.* (1979)	4.5	0.314	3.25 ± 0.17
		0.805	3.60 ± 0.18
Bateman *et al.* (1971)	4.9	3.0	$1.0 \pm 3\%$
Basu and Sinha (1956/57)	16.5	0.30	2.89 ± 0.1
De *et al.* (1972b)	16.5	1.131	2.32 ± 0.2

Note: *New standard. For references see Grieder (2001).[95]

east–west effects, longitude-dependent magnetic anomalies are also observed that manifest themselves in the muon intensities.[95]

A similar compilation as shown in Table 5 but for differential muon intensities, measured at different geographic locations, is presented in Table 6.

As mentioned earlier, because of the great wealth of existing high quality muon data in the low and mid energy (momentum) range ($\sim 0.3 \leq E \leq \sim 10,000\,\text{GeV}$) that offer an excellent data base for theoretical and experimental studies on muon propagation and energy losses in a wide range of target media, recent muon research focuses mainly on the establishment of the high and UHE region of the spectrum, for which data are scarce or non-existent.

For the absolute *vertical differential intensity*, I_v, of muons at 1 GeV/c, Allkofer *et al.* (1970a,b)[100, 101] specify the following widely used value:

$$I_v(1\,\mathrm{GeV/c}) = 3.09 \cdot 10^{-3}\ [\mathrm{cm}^{-2}\mathrm{s}^{-1}\mathrm{sr}^{-1}(\mathrm{GeV/c})^{-1}]. \tag{53}$$

For the absolute *vertical integral intensity*, J_v, Allkofer *et al.* (1975)[108] specify a value of

$$J_v(> 0.35\,\mathrm{GeV/c}) = (0.94 \pm 0.05) \cdot 10^{-2}\ [\mathrm{cm}^{-2}\mathrm{s}^{-1}\mathrm{sr}^{-1}], \tag{54}$$

for the *vertical integral flux*, J_1,

$$J_1(> 0.35\,\mathrm{GeV/c}) = (1.44 \pm 0.09) \cdot 10^{-2}\ [\mathrm{cm}^{-2}\mathrm{s}^{-1}], \tag{55}$$

and for the *omnidirectional integral intensity*, J_2,

$$J_2(> 0.35\,\mathrm{GeV/c}) = (1.90 \pm 0.12) \cdot 10^{-2}\ [cm^{-2}s^{-1}]. \tag{56}$$

10.2. *Muon flux and energy spectra*

In the following, we present a selection of atmospheric muon data recorded at or near sea level in tabulated and graphic form that may be useful for carrying out calculations as outlined above.

In Tables 7 to 9 we list as examples of differential and integral muon intensities the data obtained from measurements carried out with two different solid iron magnetic spectrographs. The data in Table 7 are from the work of Rastin *et al.* (1984a)[109] using the spectrograph at Nottingham (GB), those in Tables 8 and 9 are from the work of Matsuno *et al.* (1984)[94] using the near horizontal MUTRON spectrograph at the University of Tokyo.

The Rastin data are also plotted in differential and integral form in Fig. 16. These data include in addition some high-energy data points obtained with the Magnetic Automated Research Spectrograph at Durham (MARS), England,[110] and one set of horizontal data ($\theta = 89°$) from the work of Komori *et al.* (1977)[111] using the MUTRON detector in Tokyo are also included. Figure 17 shows the combined integral momentum spectrum assembled from the

Table 7. Best-fit vertical muon differential and integral spectra at sea level.[109]

Muon momentum [GeV/c]	Differential intensity [cm^{-2} s^{-1}sr^{-1}(GeV/c)$^{-1}$]	Integral intensity [cm^{-2} s^{-1} sr^{-1}]
0.35	$2.85 \cdot 10^{-3}$	$9.13 \cdot 10^{-3}$
0.40	$2.90 \cdot 10^{-3}$	$8.98 \cdot 10^{-3}$
0.50	$2.94 \cdot 10^{-3}$	$8.69 \cdot 10^{-3}$
0.60	$2.92 \cdot 10^{-3}$	$8.40 \cdot 10^{-3}$
0.70	$2.87 \cdot 10^{-3}$	$8.11 \cdot 10^{-3}$
0.80	$2.80 \cdot 10^{-3}$	$7.83 \cdot 10^{-3}$
0.90	$2.71 \cdot 10^{-3}$	$7.55 \cdot 10^{-3}$
1.0	$2.62 \cdot 10^{-3}$	$7.29 \cdot 10^{-3}$
1.5	$2.12 \cdot 10^{-3}$	$6.10 \cdot 10^{-3}$
2.0	$1.69 \cdot 10^{-3}$	$5.16 \cdot 10^{-3}$
3.0	$1.10 \cdot 10^{-3}$	$3.80 \cdot 10^{-3}$
4.0	$7.40 \cdot 10^{-4}$	$2.90 \cdot 10^{-3}$
5.0	$5.17 \cdot 10^{-4}$	$2.27 \cdot 10^{-3}$
6.0	$3.75 \cdot 10^{-4}$	$1.83 \cdot 10^{-3}$
7.0	$2.80 \cdot 10^{-4}$	$1.51 \cdot 10^{-3}$
8.0	$2.16 \cdot 10^{-4}$	$1.25 \cdot 10^{-3}$
9.0	$1.69 \cdot 10^{-4}$	$1.06 \cdot 10^{-3}$
10	$1.35 \cdot 10^{-4}$	$9.05 \cdot 10^{-4}$
15	$5.28 \cdot 10^{-5}$	$4.79 \cdot 10^{-4}$
20	$2.58 \cdot 10^{-5}$	$2.93 \cdot 10^{-4}$
25	$1.45 \cdot 10^{-5}$	$1.96 \cdot 10^{-4}$
30	$8.69 \cdot 10^{-6}$	$1.40 \cdot 10^{-4}$
40	$3.90 \cdot 10^{-6}$	$8.23 \cdot 10^{-5}$
50	$2.11 \cdot 10^{-6}$	$5.35 \cdot 10^{-5}$
60	$1.26 \cdot 10^{-6}$	$3.72 \cdot 10^{-5}$
70	$8.03 \cdot 10^{-7}$	$2.72 \cdot 10^{-5}$
80	$5.42 \cdot 10^{-7}$	$2.06 \cdot 10^{-5}$
90	$3.81 \cdot 10^{-7}$	$1.60 \cdot 10^{-5}$
100	$2.77 \cdot 10^{-7}$	$1.28 \cdot 10^{-5}$
150	$7.85 \cdot 10^{-8}$	$5.18 \cdot 10^{-6}$
200	$3.12 \cdot 10^{-8}$	$2.67 \cdot 10^{-6}$
250	$1.50 \cdot 10^{-8}$	$1.58 \cdot 10^{-6}$
300	$8.20 \cdot 10^{-9}$	$1.02 \cdot 10^{-6}$
400	$3.11 \cdot 10^{-9}$	$5.08 \cdot 10^{-7}$
500	$1.45 \cdot 10^{-9}$	$2.93 \cdot 10^{-7}$
600	$7.75 \cdot 10^{-10}$	$1.87 \cdot 10^{-7}$
700	$4.55 \cdot 10^{-10}$	$1.27 \cdot 10^{-7}$
800	$2.86 \cdot 10^{-10}$	$9.07 \cdot 10^{-8}$

(*Continued*)

Table 7. (Continued)

Muon momentum [GeV/c]	Differential intensity [cm^{-2} s^{-1}sr^{-1}(GeV/c)$^{-1}$]	Integral intensity [cm^{-2} s^{-1} sr^{-1}]
900	$1.89 \cdot 10^{-10}$	$6.74 \cdot 10^{-8}$
1,000	$1.31 \cdot 10^{-10}$	$5.16 \cdot 10^{-8}$
1,500	$3.14 \cdot 10^{-11}$	$1.84 \cdot 10^{-8}$
2,000	$1.13 \cdot 10^{-11}$	$8.81 \cdot 10^{-9}$
2,500	$5.11 \cdot 10^{-12}$	$4.97 \cdot 10^{-9}$
3,000	$2.67 \cdot 10^{-12}$	$3.11 \cdot 10^{-9}$

Table 8. Differential momentum spectrum of muons at zenith angle $\theta = 89°$ (MUTRON, fully corrected, Ref. 94).

Range [GeV/c]	Mean [GeV/c]	Muon number	Differential intensity $\left[\frac{1}{cm^2\,s\,sr\,(GeV/c)}\right]$	Statistical error
100–126	112	70,142	$5.69 \cdot 10^{-8}$	$\pm 2.15 \cdot 10^{-10}$
126–158	141	60,560	$3.99 \cdot 10^{-8}$	$\pm 1.62 \cdot 10^{-10}$
158–200	177	50,136	$2.41 \cdot 10^{-8}$	$\pm 1.08 \cdot 10^{-10}$
200–251	223	40,643	$1.60 \cdot 10^{-8}$	$\pm 7.94 \cdot 10^{-11}$
251–316	281	31,519	$9.88 \cdot 10^{-9}$	$\pm 5.57 \cdot 10^{-11}$
316–398	354	23,608	$5.90 \cdot 10^{-9}$	$\pm 3.84 \cdot 10^{-11}$
398–501	444	16,946	$3.37 \cdot 10^{-9}$	$\pm 2.59 \cdot 10^{-11}$
501–631	559	11,765	$1.88 \cdot 10^{-9}$	$\pm 1.73 \cdot 10^{-11}$
631–794	704	7,893	$1.02 \cdot 10^{-9}$	$\pm 1.15 \cdot 10^{-11}$
794–1,000	886	5,628	$5.29 \cdot 10^{-10}$	$\pm 7.05 \cdot 10^{-12}$
1,000–1,259	1,115	3,114	$2.71 \cdot 10^{-10}$	$\pm 4.86 \cdot 10^{-12}$
1,259–1,585	1,403	1,858	$1.31 \cdot 10^{-10}$	$\pm 3.04 \cdot 10^{-12}$
1,585–1,995	1,766	1,104	$6.23 \cdot 10^{-11}$	$\pm 1.88 \cdot 10^{-12}$
1,995–2,512	2,222	646	$2.97 \cdot 10^{-11}$	$\pm 1.17 \cdot 10^{-12}$
2,512–3,162	2,797	350	$1.29 \cdot 10^{-11}$	$\pm 6.90 \cdot 10^{-13}$
3,162–3,981	3,520	221	$6.62 \cdot 10^{-12}$	$\pm 4.45 \cdot 10^{-13}$
3,981–5,012	4,431	135	$3.25 \cdot 10^{-12}$	$\pm 2.80 \cdot 10^{-13}$
5,012–6,310	5,576	81	$1.59 \cdot 10^{-12}$	$\pm 1.77 \cdot 10^{-13}$
6,310–7,943	7,018	47	$7.33 \cdot 10^{-13}$	$\pm 1.07 \cdot 10^{-13}$
7,943–10,000	8,832	31	$3.76 \cdot 10^{-13}$	$\pm 6.75 \cdot 10^{-14}$
10,000–12,589	11,116	13	$1.18 \cdot 10^{-13}$	$\pm 3.27 \cdot 10^{-14}$
12,589–15,849	13,990	8	$5.18 \cdot 10^{-14}$	$\pm 1.83 \cdot 10^{-14}$
15,849–19,953	17,606	7	$3.50 \cdot 10^{-14}$	$\pm 1.32 \cdot 10^{-14}$
19,953–25,119	22,162	4	$1.10 \cdot 10^{-14}$	$\pm 5.50 \cdot 10^{-15}$

Table 9. Integral muon momentum spectrum (Mutron, Ref. 94)

Momentum [TeV/c]	Muon number	Integral intensity [cm^{-2} s^{-1} sr^{-1}]	Statistical error
0.10	326,499	$6.90 \cdot 10^{-6}$	$\pm 1.21 \cdot 10^{-8}$
0.13	256,357	$5.42 \cdot 10^{-6}$	$\pm 1.07 \cdot 10^{-8}$
0.16	195,797	$4.15 \cdot 10^{-6}$	$\pm 9.55 \cdot 10^{-8}$
0.20	145,661	$3.08 \cdot 10^{-6}$	$\pm 8.07 \cdot 10^{-9}$
0.25	105,018	$2.22 \cdot 10^{-6}$	$\pm 6.85 \cdot 10^{-9}$
0.32	73,499	$1.55 \cdot 10^{-6}$	$\pm 5.72 \cdot 10^{-9}$
0.40	49,891	$1.05 \cdot 10^{-6}$	$\pm 4.70 \cdot 10^{-9}$
0.50	32,945	$6.95 \cdot 10^{-7}$	$\pm 3.83 \cdot 10^{-9}$
0.63	21,180	$4.46 \cdot 10^{-7}$	$\pm 3.06 \cdot 10^{-9}$
0.79	13,287	$2.79 \cdot 10^{-7}$	$\pm 2.42 \cdot 10^{-9}$
1.00	7,659	$1.70 \cdot 10^{-7}$	$\pm 1.94 \cdot 10^{-9}$
1.26	4,545	$1.00 \cdot 10^{-7}$	$\pm 1.48 \cdot 10^{-9}$
1.59	2,687	$5.87 \cdot 10^{-8}$	$\pm 1.14 \cdot 10^{-9}$
2.00	1,583	$3.40 \cdot 10^{-8}$	$\pm 8.62 \cdot 10^{-10}$
2.51	937	$1.96 \cdot 10^{-8}$	$\pm 6.51 \cdot 10^{-10}$
3.16	587	$1.18 \cdot 10^{-8}$	$\pm 4.97 \cdot 10^{-10}$
3.98	366	$7.00 \cdot 10^{-9}$	$\pm 3.80 \cdot 10^{-10}$
5.01	231	$4.04 \cdot 10^{-9}$	$\pm 2.82 \cdot 10^{-10}$
6.31	150	$2.31 \cdot 10^{-9}$	$\pm 2.08 \cdot 10^{-10}$
7.94	103	$1.32 \cdot 10^{-9}$	$\pm 1.51 \cdot 10^{-10}$
10.00	72	$6.97 \cdot 10^{-10}$	$\pm 8.43 \cdot 10^{-11}$
12.59	59	$3.88 \cdot 10^{-10}$	$\pm 6.97 \cdot 10^{-11}$
15.85	51	$2.14 \cdot 10^{-10}$	$\pm 4.99 \cdot 10^{-11}$
19.95	44	$1.17 \cdot 10^{-10}$	$\pm 3.31 \cdot 10^{-11}$
25.12	40	$6.44 \cdot 10^{-11}$	$\pm 2.26 \cdot 10^{-11}$

MUTRON data, and data points from the DEIS spectrometer of Allkofer *et al.* (1971b)[103] in Tel Aviv and the Fréjus experiment[112] in France. A predicted spectrum from a calculation by Gaisser (1990)[113] is also plotted for comparison.

The differential muon momentum spectrum obtained from the near horizontal measurements (zenith angle 89°) made with the MUTRON detector is plotted in Fig. 18 together with data from the DEIS spectrograph, and the more recent results from the work of Gettert *et al.* (1993),[114] at Karlsruhe. The latter experiment determined the horizontal muon momentum spectrum between 250 GeV/c and 15 TeV/c, covering the zenith angle range 85° ≤ θ ≤ 90°.

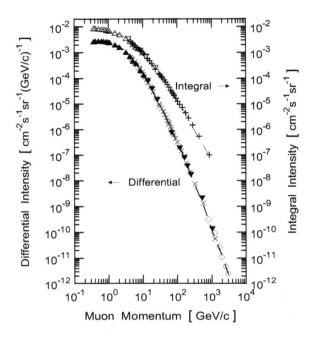

Figure 16. Differential and integral vertical muon momentum spectra determined from measurements at Nottingham (sea level).[109] Also shown are a few high momentum data points from the work of Komori *et al.* (1977),[111] Komori and Mitsui (1979), and Thompson *et al.* (1977b).[119]

▼, ▽	Appleton *et al.* (1971)[121]	▲, △	Barbouti and Rastin (1983)[108]
×, +	Rastin (1984a)[109]	○	Komori (1977)[111] and
◇	Thompson *et al.* (1977b)[120]		Komori and Mitsui (1979)[122]
− −	best fit, Rastin (1984a)[109]		

Also shown in the same plot are the data from a special spectrograph at Okayama[115] that investigated the muon spectrum at zenith angles between 70° and 78°.

In Fig. 19, we show the zenith angle (θ) dependence of the muon momentum spectrum over the full range $0° \leq \theta \leq 79°$ from

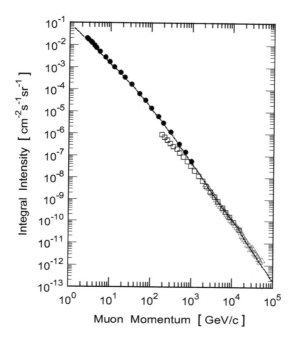

Figure 17. Combined vertical integral muon momentum spectrum at sea level in the range $1 \leq p \leq 10^5$ GeV/c. The data by Berger *et al.* (1969)[112] are derived from the Fréjus underground experiment (for details see Rhode, 1993), those from Matsuno *et al.* (1984), from the MUTRON horizontal data. The spectrum of Gaisser (1990)[113] is the result of a calculation.

•	Allkofer *et al.* (1971b)[103]	□	Matsuno *et al.* (1984)[94]
△	Berger *et al.* (1989)[112]	- - -	Gaisser (1990),[113] Theory

measurements made by Carstensen *et al.* (1978),[116] and Allkofer *et al.* (1978a, 1978b, and 1979).[117–119] Note that the ordinate is multiplied by the momentum, p, to the third power for more favorable presentation in the plot. In order to separate the individual curves, they were multiplied by a scale factor, as indicated.

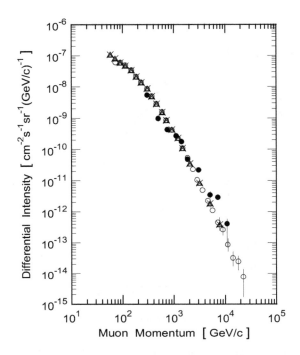

Figure 18. Comparison of the MUTRON and DEIS differential muon momentum spectra for the zenith angle range $87° \leq \theta \leq 90°$ at sea level (Refs. 123 and 124, respectively). Also shown is the more recent Karlsruhe spectrum for $85° \leq \theta \leq 90°$,[114] and the Okayama data covering the zenith angle range between $70°$ and $78°$.[115]

× DEIS ○ MUTRON • Karlsruhe △ Okayama

The next plot, Fig. 20, shows the muon charge ratio for vertically incident muons as a function of muon momentum, and the final plot, Fig. 21, shows the altitude dependence of the integral muon flux $\geq 0.3 \, \text{GeV/c}$.

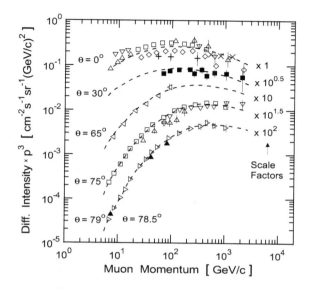

Figure 19. Differential momentum spectra of muons at sea level for different zenith angles, θ. The ordinate is multiplied by the momentum to the third power to compress the spectra. In addition, the spectra had been divided by the scale factors, as indicated, to shift the successive curves downward for better presentation. Thus, the reading at the ordinate for a given point on a curve must be multiplied by the appropriate scale factor to get the true intensity.[115,117,118,124]

	∇	Allkofer *et al.* (1971a, b)[103, 126]
Zenith Angle	$\triangle$	Nandi and Sinha (1970, 1972b)
	$+$	Abdel-Monem *et al.* (1973)
$\theta = 0°$	$\diamond$	Burnett *et al.* (1973a, b)[127,128]
	$\square$	Whalley (1974)
	$\times$	Baxendale *et al.* (1975a, b)
$\theta = 30°$	$\blacksquare$	Leipuner *et al.* (1973)
$\theta = 65°$	$\triangleleft$	Abdel-Monem *et al.* (1975)
	$\square$	Carstensen (1978),[116] Allkofer (1978b, c)[118,125]
$\theta = 75°$	$\triangle$	Asbury *et al.* (1970)
	∇	Leipuner *et al.* (1973)
$\theta = 78.75°$	$\blacktriangle$	Ashton and Wolfendale (1963)
$\theta = 79°$	$\triangleright$	Carstensen (1978),[116] Allkofer (1978b,c)[118,125]
	- - -	Maeda, calculation (1973)

Note: For references see Grieder (2001).[95]

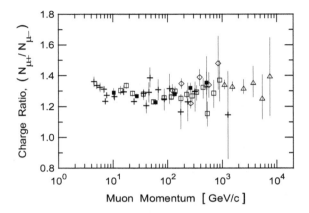

Figure 20.   Compilation of muon charge ratio data versus momentum for vertical direction. With the exception of the data by Ashley *et al.* (1975),[129, 130] that are from the Utah underground experiment located at an altitude of about 1,500 m, the data apply to sea level. The latter represent essentially the situation at sea level and were added to extend the scope to higher momenta.

■   Allkofer *et al.* (1978a)[117]   △   Ashley *et al.* (1975a,b)[129, 130]
□   Burnett *et al.* (1973a,b)[127, 128]   +   Rastin (1984b)[131]
◇   Thompson *et al.* (1977b)[120]

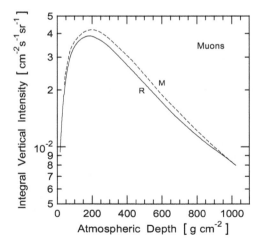

Figure 21.   Muon integral intensity as a function of atmospheric depth for muons of energy ≥ 0.3 GeV. The solid curve R is the old Rossi curve,[99] the dashed curve is from a calculation of Murakami *et al.* (1979).[131]

References

1. G. Pfotzer, *Zeitschr iff fur Physik*, **102**, 23, (1936a).
2. G. Pfotzer, *Zeitschr iff fur Physik*, **102**, 41, (1936b).
3. R. Beck, Galactic and extragalactic magnetic fields. *Space Science Review*, **99**, 243–260, (2001).
4. P. Auger, *et al.* Analyse du rayonnment cosmique á l'altitude de 3500 m (Laboratoire International du Jungfraujoch). *Journal de Physique et Radium*, **7**, 58, (1936) (in French).
5. P. Auger, *et al.* Grandes gerbes cosmique atmosphériques contenant des corpuscules ultrapénétrants. *Comptes Rendus Academic Science Paris*, **206**, 1721 (1938) (in French).
6. P. Auger, *Comptes Rendus Academic Science Paris*, **207**, 907, (1938) (in French).
7. W. Kohlhörster, I. Matthes, and E. Weber, *Naturwissenschaften*, **26**, 576 (1938) (in German).
8. L. Jánossy and A.B.C. Lovell, *Nature*, **142**, 716, (1938).
9. G.B. Khristiansen, *Cosmic Rays of Superhigh Energies*, Verlag Karl Thiemig, München (1980).
10. P.K.F. Grieder, Extensive air showers, high energy phenomena and astrophysical aspects: A tutorial, *Reference Manual and Data Book*, **2**, 1113, 681 Illustrations. ISNB: 978-3-540-76940-8; e-ISBN 978-3-540-76941-5; DOI 10.1007/978-3-540-76941-5 Springer Heidelberg, Dordrecht, London, New York (2010). Corrected 2nd Printing (2010)[f].
11. J.J. Hernàndez, *et al. Physics Letter*, 239, 1, (1990).
12. C. Caso, *et al. European Physics Journal*, **C 3**, 1, (1998).
13. F. Abe, *et al.* CDF Collaboration. *Physical Review D*, **50**, 5550, (1994).
14. H.H. Mielke, M. Föller, J. Engler, and J. Knapp, *Journal of Physics*, **G 20**, 637, (1994).
15. T. Stanev, Rapporteur Paper XXVI Internat. Cosmic Ray Conf. (1999), *AIP Conf. Proc.*, 516, 247 (2000).
16. G.B. Yodh, *et al. Physical Review D*, **27**, 1183, (1983).
17. T.K. Gaisser, *et al.*, Hadron-cross-sections-at-ultrahigh-energies-and-unitarity-bounds-on diffraction. *Physical Review*, **D 36**, 1350, (1987b).
18. M. Honda, *et al. Physical Review Letter*, **70**, 525, (1993).

[f]In the first printing, when transferring the figures from the faultless manuscript to the Springer book format, the typesetters manually redrew many of the figures and introduced countless errors, making the 1st print useless. Springer offers free exchange for 2nd printing.

19. R.M. Baltrusaitis, G.L. Cassiday, J.W. Elbert, P.R. Gerhardy, S. Ko, E.C. Loh, Y. Mizumoto, P. Sokolsky, and D. Steck, *Physical Review Letter*, **52**, 1380, (1984).

20. M. Aglietta, *et al. Astroparticle Physics*, **10**, 1 (1999).

21. N.N. Kalmykov, *et al.*, Quark-gluon-string model and EAS simulation problems at ultrahigh energies. *Nuclear Physics* B (Proc. Suppl.) **52 B**, 17–28 (1997).

22. G.J. Alner, *et al.* A general study of Proton-antiproton physics at $\sqrt{s} = 546$ GeV. *Physics Report* **154**, 247–383, (1987). UA5.

23. P.K.F. Grieder, *Nuovo Cimento*, **7 A**, 867, (1972).

24. P.K.F. Grieder, *Rivista del Nuovo Cimento*, **7**, 1, (1977).

25. K.A. Olive, *et al.* Particle physics booklet, particle data group (available from LBNL and CERN), and *Chinese Physics C*, **38**, 090001 (2014).

26. H.A. Bethe and W. Heitler, On the stopping of fast particles and on the creation of positive electrons. *Proceedings Royal Society*, London, Ser. **A 146**, 83–112, (1934).

27. W. Heitler, *Quantum Theory of Radiation*, Oxford University Press (1956).

28. A.I. Akhiezer and V.B. Berestetskii, Quantum Electrodynamics, Wiley (Interscience), New York (1965).

29. G. Källén, *Quantumelectrodynamics*, Springer Verlag, New York (1972).

30. B. Rossi, *High Energy Particles*, Englewood Cliffs, N.J. (1952).

31. R.M. Sternheimer, Density effect for the ionization loss in various materials. *Physical Review*, **103**, 511–515, (1956).

32. R.M. Sternheimer and R.F. Peierls, General expression for the density effect for the ionization loss of charged particles. *Physical Review*, **B 3**, 3681–3692, (1971).

33. A.A. Petrukhin and V.V. Shestakov, The influence of the nuclear and atomic form factors on the muon Bremsstrahlung ceoss section. *Canadian Journal of Physics*, **46**(1), S377–S380, (1968).

34. S.R. Kelner and Yu.D. Kotov, Muon energy loss to pair production. *Soviet Journal Nuclear Physics*, **7**(2), 237–240, (1968a).

35. S.R. Kelner and Yu.D. Kotov, Pair production by muons in the field of nuclei. *Canadian Journal of Physics* **46**, S387–S390, (1968b).

36. R.P. Kokoulin and A.A. Petrukhin, Influence of the nuclear form factor of the cross section of electron pair production by HE muons. *PICRC*, **6**, 2436–2444, (1971).

37. A.G. Wright, A critical evaluation of theories of direct electron pair production by muons. *Journal of Physics A*, **6**, 79–92, (1973).

38. T. Murota and A. Ueda, On the foundation and application of Williams-Weizsäcker method. *Progress of Theoretical Physics*, **16**, 497–506, (1956).

39. D. Kessler and P. Kessler, On the validity of the Williams-Weizsäcker method and the problem of the nuclear interactions of relativistic μ-mesons. *Nuovo Cimento*, **4**, 601–609, (1956).

40. D. Kessler and P. Kessler, Sur une méthode simplifiée de calcul pour les processus relativistes en electrodynamique quantique. *Nuovo Cimento*, **17**, 809–827, (1960).

41. K. Daiyasu, *et al.*, On hard showers produced by μ-mesons. *Journal Physical Society Japan*, **17**, Suppl., A-III, 344–347, (1962).

42. K. Kobayakawa, Fluctuations and nuclear interactions in th energy loss of cosmic-ray muons. *Nuovo Cimento*, **47B**, 156–174, (1967).

43. K. Kobayakawa and S. Miono, Application of the birth-and-death processes to the cascade theory. *Canadian Journal of Physics*, **46**, S212–S215, (1968).

44. K. Kobayakawa, On the range fluctuations of high energy muons. *PICRC*, **5**, 3156–3161, (1973).

45. L.B. Bezrukov, *et al.*, Investigation of nuclear interactions of cosmic ray muons underground. *PICRC*, **6**, 2445–2450, (1971).

46. V.V. Borog and A.A. Petrukhin, The cross section of the nuclear interaction of high energy muons. *PICRC*, **6**, 1949–1954, (1975).

47. L.B. Bezrukov and E.V. Bugaev, Nucleon shadowing effects in photon-nucleus interactions. *Yad. Physics*, **33**, 1195–1207, (1981), and *Sov. J. Nucl. Phys.*, **33**, 635, (1981).

48. J.D. Jackson and R.L. McCarthy, z^3 Corrections to energy loss range. *Physical Review*, **B 6**, 4131–4141, (1972).

49. R.P. Kokoulin and A.A. Petrukhin, Analysis of the cross section of direct pair production by fast muons. *Acta Physics Academy Science Hungarica*, **29**, S4, 277–284, (1970).

50. A. Crispin and G.N. Fowler, Density effect in the ionization energy loss of fast charged particles in matter. *Review Modern Physics*, **42**, 290–316, (1970).

51. E.V. Bugaev, *et al.* Cosmic Muons and Neutrinos. *Atomizdat, Moscow 1970*, 38 (in Russian) (1970).

52. Y.P. Kotov and V.M. Logunov, Energy losses and the absorption curve of muons. *Acta Physics Academy Science Hungarica*, **29**, S4, 73–79, (1970).

53. L. Bergamasco and P. Picchi, Muon energy losses and range straggling in the $10 < E < 10^5$ GeV region. *Nuovo Cimento*, **3 B**, 134–148, (1971).

54. Yu. N. Vavilov, *et al.*, Intensity of cosmic muons at large rock and water depths in ground and their energy spectrum at sea level. *Yad. Fiz.*, **18**, 844–853 (October 1973), and *Soviet Journal of Nuclear Physics*, **18**, 434–439, (1974).
55. C. Grupen, Electromagnetic interactions of high energy cosmic ray muons. *Fort. d. Phys.*, **23**, 127–209, (1976).
56. W. Lohmann, *et al.*, Energy loss of muons in the energy range 1–1000 GeV. CERN Yellow Report 85-03 (1985).
57. E. V. Bugaev, *et al.* Muon depth-intensity relation and data of underground and underwater experiments. *Proceedings of Nestor Workshop*, 268–305, (1993).
58. E. V. Bugaev, *et al. Università degli Studi di Firenze, Dipartimento di Fisica and Instituto Nationale di Fisica Nucleare Sezione di Firenza,* Preprint DFF 204/4/1994 (1994).
59. R. K. Adair, In Muon Physics I, Electromagnetic Interactions. Hughes, Vernon, W., and C.S. Wu, ed., Academic Press, New York (1977).
60. D. Chirkin and W. Rhode, Propagating leptons through matter with muon Monte Carlo (MMC). arXiv:hep-ph/0407075 3 Auh (2016).
61. B. Rossi and K. Greisen, Cosmic ray theory. *Review of Modern Physics*, **13**, 240–309, (1941).
62. R.M. Barnett, *et al.* Review of particle physics. *Physical Review*, **D 54**, 1–1241, (1996).
63. U. Fano, Penetration of protons, alpha particles, and mesons. *Annual Review of Nuclear Science*, **13**, 1–66 (1963).
64. R.M. Sternheimer *et al.* The density effect for the ionization loss of charged particles in various substances, *Atomic and Nuclear Data Tables*, **30**, 261 (1984).
    See also Groom, *et al.*, *ibid* 78, 183–356, (2001) and D.Yu., Ivanov, *et al.*, *Physics Letter*, **B 442**, 453–458 (1998) for corrections. Production of e^+e^- pairs to all orders of $Z\alpha$ for collisions of high energy muons with heavy nuclei. General expression for the density effect for the ionization loss of charged particles.
65. R.F. Christy and S. Kusaka, The interaction of γ-rays with mesotrons. *Physical Review* **59**, 405–414, (1941a).
66. R.F. Christy and S. Kusaka, Burst production by mesotrons. *Physical Review*, **59**, 414–421, (1941b).
67. W.K. Sakumoto, *et al.* Measurement of TeV muon energy loss in iron. *Physical Review*, **D 45**, 3042–3050, (1992).
68. S. Eidelman, *et al.*, Particle Physics Booklet, Particle Data Group, Springer, Berlin (available from LBNL and CERN) (2004), and *Physics Letter B*, **592**, 1–1109, (2004).

69. T. Pierog *et al.* Latest results of air shower simulation programs CORSIKA and CONEX. *PICRC*, **4**, HE-1.6, Code 0899, 625, (2007).

70. P. Berghaus, Muons in IceCube, for the IceCube Collaboration: *Nuclear Physics B (Proc. Suppl.)*, **196**, 261–266, (2009).

71. S.R. Kelner, Pair production in collisions between a fast particle and a nucleus. *Yadernaya Fiz.*, **5**, 1092–1099, (1967), and *Soviet Journal of Nuclear Physics*, **5**, 778–783, (1967).

72. A.A. Petrukhin and V.V. Shestakov, Interactions of cosmic muons of high energy. *Physics of Elementary Particles, Atomizdat*, (1966).

73. M.L. Morris and R.O. Stenerson, Muon pair production by cosmic ray muons. *Nuovo Cimento*, **53 B**, 494–506, (1968).

74. J.C. Barton and I.W. Rogers, The study of the production of muon pairs by muons. *Acta Physical Academy Science Hungarica*, **29**, S4, 259–262, (1970).

75. J.J. Russell, *et al.* Observation of muon trident production in lead and the statistics of the muon. *Physical Review Letter*, **26**, 46–50, (1971).

76. C.F.W. Weizsäcker, Ausstrahlung bei Stössen sehr schneller Elektronen. *Zeitschr iff fur Physics*, **88**, 612–625, (1934).

77. E.J. Williams, Application of the method of impact parameter in collisions. *Procceedings of Royal Society*, **A 139**, 163–186, (1933).

78. E.J. Williams, Correlation of certain collision problems with radiation theory. *Kgl. Dansk. Vidensk. Selsk. Math. Fys. Medd.*, 13(1), 1–50, (1935).

79. P. Kessler and D. Kessler, Généralisation de la méthode de Williams et Weizsäcker. *Comptes Rendus*, **244**, 1896–1898, (1957).

80. S. Higashi, *et al.* Multiple pion production by high energy muons. *Nuovo Cimento*, **38**, 107–129, (1965).

81. V.V. Borog, *et al.* Study of nuclear and electromagnetic interactions of high energy muons. *Canadian Journal of Physics* **46**, S381–S386, (1968).

82. B.M. Chasan, *et al.*, Multiple meason production by photons in hydrogen. *Physical Review* **119**, 811–814, (1960).

83. H.R. Crouch, Jr., *et al.*, Gamma ray proton interactions between 0.5 and 4.8 BeV. *Physical Review Letter*, **13**, 636–639, (1964).

84. J. A. Kirk, *et al.*, Inelastic muon interactions in nuclear emulsion at 2.5 and 5.0 GeV. *Nuovo Cimento*, **40**, 523–541, (1965).

85. G.N. Fowler and A.W. Wolfendale, The interaction of mu-mesons with matter. *Progress in Elementary Particle and Cosmic Ray Physics*, **4**, 107, (1958).

86. D.P. Bhattacharyya, Calculation of muon range spectrum under seawater from the latest JACEE primary spectrum using the modified

energy loss formulation after Kobayakawa. *Nuovo Cimento*, **9 C**, 404–413 (1986).

87. S. Chapman, The absorption and dissociative or ionizing effect of monochromatic radiation in an atmosphere on a rotating Earth: Part II. Grazing incidence. *Proceedings of the Physical Society (London)*, **43** (26), 483–501, (1931).

88. J. A. Fitzmaurice, Simplification of the Chapman function for atmospheric attenuation. *Applied Optics*, **3**, 640–640, (1964).

89. W. Swider and M.E. Gardner, Environmental Research Papers No 272, Air Force Cambridge Research, Bedford, MA (1967).

90. Jr. S. William, The determination of the optical depth at large solar zenith distances. *Planetary and Space Science*, **12**, 761–782, (1964).

91. G. Brasseur and S. Solomon, Aeronomy of the Moddle Atmosphere, D. Reidel Publishing Company, Dordrecht/Boston/Lancaster (1986).

92. C.A. Ayre, *et al.* The Durham 5000 GeV/c spectrograph MARS. *PICRC*, **4**, 547–551 (1969a).

93. C.A. Ayre, *et al.* The Durham 5000 GeV/c spectrograph MARS. *Nuclear Instruction Methods*, **69**, 106, (1969b).

94. S. Matsuno, *et al.*, Cosmic ray muon spectrum up to 20 TeV at 89° zenith angle. *Physical Review* **D 29**, 1–23, (1984).

95. P.K.F. Grieder, Cosmic Rays at Earth (Researcher's Reference Manual and Data Book), 1103 pages, Elsevier Science B.V., Amsterdam, The Netherlands, ISBN 0444507108 (2001).

96. R.I. Enikeev, *et al.*, Study of inelastic interactions of muons with NaCl nuclei in energy range up to 4 TeV. *PICRC*, **7**, 82–85 (1983).

97. O. C. Allkofer, *et al. Canadian Journal of Physics* **46**, S301, (1968).

98. K.S. Chandrasekharan, *et al. Proceedings of the Indian Academy of Sciences*, **32**, 95, (1950).

99. B. Rossi, *Review of Modern Physics*, **20**, 537 (1948).

100. O.C. Allkofer, *et al. Physics Letter*, **31** B, 606 (1970a).

101. O.C. Allkofer, *et al. Proc. VI. Inter-American Seminar on Cosmic Rays*, La Paz, IV, 937 (1970b).

102. O.C. Allkofer, *et al. Physics Letter*, **36** B, Erratum, 428 (1971d).

103. O.C. Allkofer, *et al. Physics Letter*, **36** B, 425 (1971b).

104. O.C. Allkofer, *et al. PICRC*, **4**, 1314, (1971c).

105. O.C. Allokfer and K. Clausen, *Acta Physical Academic of Science Hungary*, **29** (2) 689, (1970).

106. O.C. Allkofer, University of Kiel, Internal Report (1965).

107. A I Barbouti and B C Rastin, A study of the absolute intensity of muons at sea level and under various thicknesses of absorber, 1983 *J. Phys. G: Nucl. Phys.* 9 1577

108. O.C. Allkofer, *et al. Journal of Physics* **G 1**, L51, (1975a).

109. B.C. Rastin, *Journal of Physics G* **10**, 1609 (1984a).

110. M.G. Thompson, *et al. Journal of Physics G* **3**(2), L39, (1977a).

111. H. Komori, *PICRC*, **6**, 26, (1977).

112. Ch. Berger, *et al.* Fréjus Collaboration, *Physics Letter*, **227 B**, 489, (1989).

113. T.K. Gaisser, *Cosmic Rays and Particle Physics*, Cambridge (1990).

114. M.J. Gettert, *et al. PICRC*, **4**, 394 (1993).

115. S.T. Tsuji, *et al. PICRC*, **1**, 614, (1995).

116. K. Carstensen, Ph. D. Thesis, University of Kiel (1978).

117. O.C. Allkofer, *et al. Physical Review Letter*, **41**, 832 (1978a).

118. O.C. Allkofer, *et al. Proc. VI. European Cosmic Ray Symposium KIEL*, 72, and University of Kiel Report IFKKI 78/3 (1978b).

119. O.C. Allkofer, *et al. PICRC*, **10**, 50 (1979).

120. M.G. Thompson, *et al. PICRC*, **6**, 21, (1977b).

121. I.C. Appleton, M.T. Hogue, and B.C. Rastin, *Nuclear Physics*, **B 26**, 365 (1971).

122. H. Komori and K. Mitsui, *PICRC*, **10**, 65, (1979).

123. T. Kitamura, *PICRC*, **13**, 361, (1981).

124. O.C. Allkofer, *et al. PICRC*, 6, 38, (1977).

125. O.C. Allkofer, *et al. Proc. VI. European Cosmic Ray Symposium KIEL*, 71 (1978c).

126. O.C. Allkofer, K. Carstensen, and W.D. Dau, *Physics Letter*, **36** B, 425, (1971a).

127. T.H. Burnett, G.E. Masek, T. Maung, E.S. Miller, H. Ruderman, and W. Vernon, *PICRC*, **3**, 1764 (1973a).

128. T.H. Burnett, L.J. LaMay, G.E. Masek, T. Maung, E.S. Miller, H. Ruderman, and W. Vernon, *Physical Review Letter*, **30**, 937, (1973b).

129. G.K. Ashley, *et al. PICRC*, **12**, 4282 (1975a).

130. G.K. Ashley, *et al. Physical Review D*, **12**, 20 (1975b).

131. B.C. Rastin, *Journal of Physics*, **G 10**, 1629 (1984b).

132. K. Murakami, K. Nagashima, S. Sagisaka, Y. Mishima, and A. Inoue, *Nuovo Cimento*, **2C**, 635 (1979).

133. P. Amaral, *et al.*, ATLAS TileCal Collaboration: *European Physics Journal*, **C 20**, 487–495 (2001). A precise measurement of 180 GeV muon energy loss in iron.

134. M. Antonelli, *et al. Proc. 6th Internat. Conf. on Calorimetry in High Energy Physics*, 1996, Frascati, Italy, 561 (1996), and Design Report, CERN/LHCC 96-40, p. 150 (1997).

135. G. Bonomi, *et al.*, Applications of cosmic ray muons. *Progress in Particle Nuclear Physics*, **112**, 103768, (2020).

136. S. Chin, *et al.*, A study of nuclear interactions of muons with energy higher than a few 100 GeV. *PICRC*, **3**, 1971–1976 (1973).

137. S.R. Kelner, *et al.* About cross section for high energy muon Bremsstrahlung. preprint 024-95, Moscow Technical University (1995).

138. S.R. Kelner, *et al.* Bremsstrahlung from muons scattered by atomic electrons. *Physics Atomic Nuclei*, **60**(4), 576–583, (1997), and *Yad. Fiz.* **60**, 657–665 (1997).

139. S.R. Kelner, Pair production in collisions between muons and atomic electrons. *Physics of Atomic Nuclei*, **61**, 448–456, (1998).

140. S.R. Kelner, and A.M. Fedotov, Bremsstrahlung from muons scattered by atomic electrons. *Physics of Atomic Nuclei*, **62**, 307, (1999a).

141. K. Mitsui, Muon energy loss distribution and its application to muon energy determination. *Phys. Rev.* D, **45**, 3051–3061 (1992).

142. M.J. Tannenbaum, Report CERN-PPE/91-134, 19 August 1991. Comparison of two Formulas for Muon Bremsstrahlung.

143. G. Battistoni, *et al.* Study of radiative interactions of 300 GeV muons in a high resolution calorimeter. *PICRC*, 1, 597–600, (1995).

144. M.G. Aartsen, *et al.* Characterization of the atmospheric muon fluc in IceCube. arXiv:1506.07981 [astro-ph.HE] 16 Mar (2016).

145. A. Sandrock, *et al.* Radiative corrections to the average Bremsstrahlung energy loss of high energy muons. *Physics Letter*, **B 776**, 350–354, (2018).

146. C. Spiering, Proc. Workshop on "Simulation and analysis methods for large nutrino telescopes", Zeuthen, Germany, July 6-9, 1998, C. Spiering (ed). Conf. C08-07-06.8, DESY-PROC-1999-01 (499) (1999).

147. T. Murota *et al.* The creation of an electron pair by a fast charged particle, *Progress of Theoretical Physics*, **16**, 482–496, (1956).

148. A.I. Barbouti, and B.C. Rastin, *Journal of Physics*, **G 9**, 1577, (1983).

149. Rhode *et al.*, Limits on the flux of very high energy neutrinos with the Fréjus detector, Astroparticle Physics Volume 4, Issue 3, February 1996, Pages 217–225

https://doi.org/10.1142/9789811264917_0003

Chapter 3

Principle of Cosmic Muography — Techniques and Review

Paolo Checchia

*INFN sezione di Padova,
via F. Marzolo 8 I 35131 Padova, Italy
Paolo.checchia@pd.infn.it*

This chapter reviews the muon interaction with matter and the consequent use in cosmic muography. On the basis of the object to be studied, the physical process, and the experimental setup, different techniques based on muon transmission, absorption, multiple Coulomb scattering (MCS), and coincidence with fission neutrons are described and, when possible, compared. They all aim at reconstructing images of an inaccessible volume and thus a short review of the related algorithms is also given.

1. Introduction

Cosmic rays are a natural source of particles falling to the Earth's surface with a major component, at sea level, constituted by muons, that is, particles that have the characteristic of penetrating thick and heavy materials. Given the capability to detect and to measure such particles conceived in decades of high energy Physics studies, it is possible to develop techniques to exploit this natural and free source for several interdisciplinary and civil applications.

As a general definition, cosmic muon technology regards all techniques that profit from the penetration capability of cosmic muons to reconstruct the content of large and inaccessible volumes. However, for consistency with other chapters and with several contributions in literature, it is convenient to use the term cosmic muography instead.

This definition intends to include all possible definitions as found in the literature, either generic or specific ones (e.g., *muon radiography, muon tomography, muon absorption radiography, muon scattering tomography,* etc.), with possible overlaps and confusing concepts.

It has to be considered that, to investigate the content of systems which can differ by orders of magnitude in dimensions, mass, structure, etc., one can profit from different physical interactions affecting muons when they cross material, one can use different experimental setups and different data-taking procedures. Often, specific definitions are related to specific physical processes (e.g., *muon absorption radiography*) although this is not always true and, therefore, it is convenient to recall the physical process and the most common experimental setups.

In this chapter, the muon interactions with matter are first recalled (energy loss and MCS) and then, after a brief summary of the Monte Carlo simulation tools, the possibilities of installing detectors only on one side of the object to study or on both sides are distinguished. On this basis, three different techniques are described (transmission, absorption, and scattering muography) with the most relevant algorithms related to them. Furthermore, a specific technique based on the coincident detection of muons and neutrons is recalled.

2. Muon Interactions with Matter

As discussed in the previous chapter, muons are the main component of cosmic rays at sea level. Although they are unstable particles, their lifetime and energy spectrum ensure a path of several kilometers before they decay and hence this characteristic has negligible implications for cosmic muography. Being leptons, muons do not have strong interactions with atomic nuclei, although, as other charged particles, they undergo energy losses and MCS.

2.1. *Energy losses*

An exhaustive treatment of different effects contributing to muon energy losses is given in Sections 1.7 and 1.8. Here, it is sufficient to recall a few additional concepts that often occur in literature. The energy loss $-dE/dx$ is also called *mass stopping power* and the denominator dx is the product $dl \cdot \rho$ of an elementary length dl crossed by the particle in a material with density ρ. Usually, the mass stopping power is given in $\mathrm{MeV\,g^{-1}\,cm^2}$ and the quantity $x = l \cdot \rho$, referred to in the literature as *opacity*, is given in $\mathrm{g/cm^2}$. It is also worth mentioning that, in case of a compound or a mixture of elements of density ρ_c and a fraction by weight w_i of the ith element, the energy loss can be obtained as

$$\frac{dE}{dx} = \frac{1}{\rho_c}\frac{dE}{dl} = \sum_i w_i \left(\frac{dE}{dx}\right)_i = \sum_i \frac{w_i}{\rho_i}\left(\frac{dE}{dl}\right)_i. \qquad (1)$$

Figure 4 of Chapter 1 shows the muon energy loss in standard rock as a function of the muon energy. One can note that, below about $100\,\mathrm{GeV}$, where a large fraction of cosmic ray muons resides, the dominating contribution is due to ionization and atomic excitation. Rewriting Eq. (16) of Chapter 1 in a simplified form as

$$-\frac{dE}{dx} = a(E) + b(E)E, \qquad (2)$$

the first term dominates as it can be seen in Table 3 of Chapter 1.

For several applications and in particular for the cosmic muography, it is important to mention the particle *range* which quantifies how much a muon can penetrate in a given material before it loses all its energy. It can be given in opacity units ($\mathrm{g\,cm^{-2}}$) and it can be obtained by integrating the inverse of Eq. (2) as Groom *et al.* (2001)[1]

$$X(E) = \int_{E_{\min}}^{E} \left(\frac{dE'}{dx}\right)^{-1} dE' = \int_{E_{\min}}^{E} (a(E') + b(E')E')^{-1}dE', \qquad (3)$$

where $E_{\min}$ is a minimum energy value sufficient (e.g.,) to detect the particle and so small that the result is insensitive to its value. The range X for muons of different energies crossing standard rock ($\langle Z \rangle = 11$, $\langle Z/A \rangle = 0.5$, $\rho = 2.65\,\mathrm{g/cm^3}$) is shown in Table 1 where, for convenience, the linear range $X_L = X/\rho$ has also been included. In the table, the range mean values for muons of the same energy are shown, although there are fluctuations in the energy loss and hence the range of individual particles presents a distribution which is, in a

P. Checchia

Table 1. Muon range in standard rock ($\langle Z \rangle = 11, \langle Z/A \rangle = 0.5, \rho = 2.65\,\mathrm{g/cm^3}$).

E	X	X_L
GeV	$10^3\,\mathrm{g\,cm^{-2}}$	m
1	0.5	2
10	4.9	18
100	40.8	154
1,000	245.3	926

first approximation, Gaussian. This is known as *range straggling* and its fractional value is defined as the square root of the range variance over the range. The fractional straggling depends the on material and on the particle mass, and for muons in copper, it is around 2.8% to 5.7% from 300 MeV to 10 GeV.[1] Above about 100 GeV, the fluctuations in bremsstrahlung losses begin to dominate, and hard losses are more probable in bremsstrahlung than in other processes. For high energetic muons crossing an opacity X, it is convenient to define the survival probability $P_s(E, X)$, which can be handled by Monte Carlo simulations.[2] The muon range in rocks and the fluctuations at high energies are important for transmission muography as described in more details in what follows.

2.2. *Multiple Coulomb scattering*

As any other charged particle, cosmic-ray muons when crossing a medium are deflected by many small angle elastic Coulomb scatterings from nuclei and only occasionally by single large angle (hard) collisions. When the crossed thicknesses are not small, as in cosmic muon applications, one can follow the MCS approximation to describe the process[3] as illustrated in Fig. 1. The distribution of the scattering angle projected onto a plane, for charged particles of momentum p, can be approximated by a Gaussian with a standard deviation given by

$$\sigma_\theta = \frac{13.6\,MeV}{\beta c p} z \sqrt{\frac{x}{X_0}} \left[1 + 0.038 \ln \left(\frac{x z^2}{X_0 \beta^2} \right) \right], \qquad (4)$$

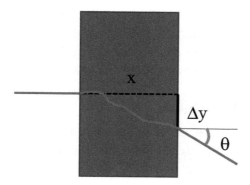

Figure 1. Basic description of multiple Coulomb scattering of a charged particle (red lines) projected onto a plane.

where βc and z are the velocity and the charge number of the incident particle, respectively, x is the thickness of the crossed medium, and the radiation length X_0 is given by

$$X_0 = 716.4 \, \text{g/cm}^3 \frac{A}{Z(Z+1)\ln\left(\frac{287}{\sqrt{Z}}\right)}. \qquad (5)$$

The scattering in the space can be obtained from

$$\sigma_\theta = \frac{1}{\sqrt{2}} \theta_{space}^{rms}.$$

In Eqs. (4) and (5), both medium thickness and radiation length are given as opacities. Since in Eq. (4) the standard deviation depends on their ratio $t = x/X_0$, the relation holds also for linear quantities. It is relevant to note that, for a sample of non-monochromatic particles, as cosmic-ray muons, the scattering angular distribution projected onto a plane is by far not Gaussian. The presence of the factor $1/p$ indicates that angular distribution of cosmic muons crossing a medium of a given thickness is approximately a convolution of Gaussian distributions

$$\frac{dN}{d\theta} = \frac{1}{k\sqrt{2\pi t}} \int_{p_{min}}^{\infty} dp \, p \, f(p) \, e^{\frac{-p^2\theta^2}{2tk^2}} \qquad (6)$$

where $k = 13.6/(\beta c)\,\text{MeV}$, $f(p)$ is the muon momentum spectrum and p_{min} is the minimum momentum of muons that are not absorbed.

In Eq. (6), the logarithmic term in t of Eq. (4) has been neglected. The momentum convolution effect is much more relevant in producing long tails in the angular distribution than the approximation that does not take into account single hard scattering.

3. Muography Techniques

This section focuses first on the three main techniques that profit from the cosmic muon interaction with matter described above to study the content of the object of interest. The two first ones, namely transmission and absorption techniques, are based on energy losses and are separated on the basis of the experimental setup.[4] The third one exploits MCS of charged particles in a medium to provide information on its structure. In addition, a fourth technique is described, which can be fundamental for detecting fissile materials by requiring a coincident detection of muons and neutrons. All techniques profit from a natural source that is free and safe compared with any other artificial source. However, since cosmic ray flux intensity is quite low, being compatible with life on Earth, they all require quite a long time to collect sufficient data.

It is worth remembering that the study of possible applications of cosmic muography as well as the development of algorithms to optimize data analysis and image reconstruction can require large and expensive detectors and targets which are not easily available. Therefore, to study a possible application it may be necessary to rely on Monte Carlo simulation tools (i) to generate the cosmic-ray muons, (ii) to describe precisely the system (detectors and target) crossed by muons, and (iii) to track particles through the whole setup. The last two items are common practice in Particle Physics and one can profit of a simulation toolkit, GEANT4,[5] to implement detailed geometry and material description. The package contains all the physics processes affecting particle interactions with matter. With such a package, none of the approximations proposed in the previous section are necessary and hence the description of muon behavior in the matter is as accurate as possible. On the other hand, the cosmic muon generator is often custom-made and optimized for specific applications, although there are roughly two groups of public packages for this task. In the former group the entire cascade of secondary particles started by primary cosmic rays is simulated or parameterized. Among such packages one can quote CORSIKA,[6] MCEq,[7] and CRY.[8] In the second group, more suitable for cosmic muography, only cosmic muons are generated according to parametrizations of differential energy and angular flux based on experimental data (EcoMug)[9] or on data obtained from the previous group of generators (CMSGEN).[10] In particular, the EcoMug generator, recently released, uses a default parametrization based on ADAMO data[11] as shown in Fig. 2, but it allows the user to generate muons according to any parametrization of the differential muon flux and with different geometrical generation methods. For applications where muons are expected to cross very thick objects, the relevant muon flux is limited to the high energy part of the spectrum and this may require dedicated generators.

Simulated data obtained from muon generation, followed by GEANT4 tracking, allow to study the expected results of the

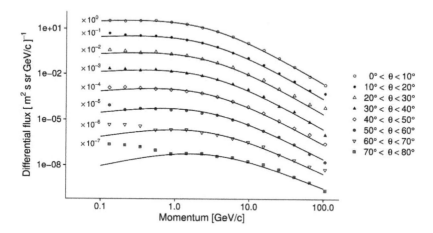

Figure 2. Experimental momentum spectrum for eight zenith angle intervals with the parametrization curves. The excess at low momenta and high zenith is attributed to the residual electron component of cosmic rays. To avoid overlaps, data and curves at different zenith angles are multiplied by the reported factors. This figure is from Ref. 9.

application with high reliability as discussed in Chapter 10. Furthermore, quantities not available in real data as the individual momentum can be used in simulated data to understand the origin of problematic effects or to optimize the detector response.

3.1. *Transmission technique*

This technique corresponds to the first application of the use of cosmic-ray muons which dates back to 1955[12] to measure *"the thickness of rock overlying an underground powerhouse"* in Australia and to 1970[13] to investigate the interior of Chephren's pyramid at Giza. The technique is based on the measurement of attenuation of the cosmic ray flux intensity produced by massive objects.

For the simplest applications, when one has just to determine average variations on the amount of material crossed by particles, (e.g., height of liquid in a water reservoir) it may be sufficient to count the muon rate with a stable and reliable detector pointing to the region of interest.

For more complex applications, transmission is the only possible technique that allows to examine very large volumes as volcanoes, mines, pyramids, etc. In such cases, there is no practical alternative to place one or more muon detectors downstream the object to study as depicted in Fig. 3 and to consider that the higher the opacity crossed by particles, the lower the number of muons having enough energy to cross it.

The evaluation of the muon attenuation is obtained by comparing the flux measured with tracker detectors pointing to the target object with the flux measured in absence of any obstacle from the detector and the free sky. The detector(s) should provide a good tracking performance with good angular resolution and bi-dimensional measurements. The single detector size may not be relevant (of the order of a squared meter) provided a large angular acceptance is guaranteed.

A typical detector for this technique (see the dedicated chapter for more details) might be based on two or three sensitive planes (each measuring two orthogonal coordinates) placed at a distance D between the first and the last plane. If each plane measures the muon

Figure 3. Pictorial illustration of the muon transmission technique. A detector (not in scale) placed on the flank of a mountain measures the flux of muons crossing the rock.

impact coordinates with a similar resolution d, the angular resolution of the detector is approximately $\Delta\theta \sim \Delta\varphi \sim d/D$. Given a size S for each plane, the total angular aperture is of the order of S/D and for $d \sim 5\,\text{mm}, S \sim D \sim 1\,\text{m}$ one obtains an angular resolution $\Delta\theta \sim \Delta\varphi \sim 5$ mrad and a maximum angle of 45°. Since the detector can be placed at several hundred meters from the target, the spatial resolution can be of the order of a few meters. Considering that the expected multiple scattering of muons inside the target may be of the same order as the detector angular resolution, detectors with a better resolution may be useless.

For other aspects, the muon scattering in the crossed matter can be neglected and hence a muon trajectory can be assumed to be a straight line. Considering a medium length L, the opacity is $X(L) = L\rho$ if the density is constant or, if it is not the case,

$$X(L) = \int_L \rho(x)dx = L\bar{\rho}. \tag{7}$$

It is clear from Eq. (3) that only muons with sufficient energy can emerge from the medium and then, since the spectrum of cosmic muons is continuous, the number of muons measured at a given angle depends on the crossed opacity. Given a certain detector, the number of muons measured in a certain time interval Δt can be expressed in terms of target opacity $X_T(\theta, \varphi)$, zenithal angle θ, and azimuthal angle φ as $N_T(\theta, \varphi; X_T) = \Delta t \times S_{eff}(\theta, \varphi) \times \Phi(\theta, \varphi; X_T)$. Geometrical acceptance and all efficiencies (e.g., trigger, detector, selection, etc.) are included in the term $S_{eff}(\theta, \varphi)$, while $\Phi(\theta, \varphi; X_T)$ represents the flux of cosmic-ray muons crossing the medium.

One can then define the measured transmission rate as

$$T_m(\theta, \varphi; X_T) = \frac{\Delta t_{FS}}{\Delta t_T} \frac{N_T(\theta, \varphi; X_T)}{N_{FS}(\theta, \varphi; 0)} \tag{8}$$

where N_T and N_{FS} are the numbers of muons recorded during target and free sky measurements with acquisition times Δt_T and Δt_{FS}, respectively. The sky opacity is assumed to be negligible. In principle, one could object that this assumption is too crude given that free sky cosmic-ray flux depends on several factors, including atmospheric pressure, as discussed in the previous chapter. However, variations are quite small, they concern only the low energy part of the spectrum, it is possible to monitor several external factors to introduce corrections and it is expected that the flux variations enter only as a scale factor, with little dependence on the direction angles.

Since in Eq. (8) one can consider the same S_{eff} for the two measurements, provided the detector conditions are stable, the measured transmission is directly related to the muon flux rate as follows:

$$T_m(\theta, \varphi; X_T) = \frac{\Phi_T(\theta, \varphi; X_T)}{\Phi_{FS}(\theta, \varphi; 0)}. \tag{8'}$$

The measured transmission must be compared with the expectations to determine the opacity given the measurements. In a first step, one should integrate Eq. (2) to obtain the minimum energy E^T that muons need to cross an opacity X. Then, after parametrizing the differential flux $d\Phi/dE$ as a function of muon energy, one should calculate the expected flux

$$\Phi_{TE}(\theta, \varphi; X_T) = \int_{E^T(X)}^{\infty} \frac{d\Phi}{dE} dE$$

and eventually obtain

$$T_E(\theta, \varphi; X_T) = \frac{\Phi_{TE}(\theta, \varphi; X_T)}{\Phi_{FS}(\theta, \varphi; 0)}.$$

However, in most cases the expected transmission distribution is estimated with a detailed simulation which allows to avoid approximations and simplified medium descriptions. With simulation it is also possible to obtain the opacity as a function of the transmission: $X_T = F(T)$. By comparing T_m and T_E or by looking for deviation from unity of the ratio $R(\theta, \varphi) = T_m(\theta, \varphi)/T_E(\theta, \varphi)$, one can have the medium opacity (as $X_T = F(T_m)$) or deviations from the expectations. An example of a map of $R(\alpha, \phi)$ (in this case the zenith angle has been substituted by the elevation angle α) in a search for cavities in Mt. Echia at Naples is shown in Fig. 4.[14, 15] As a last step, if the length $L(\theta, \varphi)$ of the medium crossed by muons is known or can be measured, from $X_T(\theta, \varphi)$ it is straightforward to calculate the two-dimensional (2D) density map of the average medium density $\overline{\rho}(\theta, \varphi)$.

With only one detector and with data taken from a single position, this technique provides a 2D image and density map as a function of zenith/elevation and azimuthal angles. For this reason, this technique is often called *muon radiography*. However, data taken from several positions, either simultaneously with several detectors or in sequence by moving a single tracker, can be combined to produce a 3D image of the target in terms of material density.

Since one of the first examples[16] applied to a volcano study, other applications of 3D reconstruction based on transmission muography have been published. A spectacular example of such a reconstruction

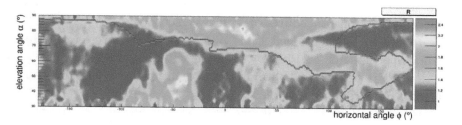

Figure 4. Example of the ratio R obtained during a study of Mt. Echia in Naples. Green regions correspond to $R > 1$ and indicate the possible presence of cavities. The expected signal shape of a known test chamber is indicated by the red line, but there is a clear evidence of the presence of additional cavities. This figure is from Ref. 15.

has been recently obtained by the — ScanPyramids collaboration that has found the indication of an unknown new chamber in Khufu's Pyramid in Egypt.[17] The discovery has been achieved by using three types of detectors, two located in the internal cavities, namely Grand Gallery and Queen's chamber, and the 3rd one outside the pyramid. A 3D reconstruction of the above-mentioned example of monte Echia is available in Cimmino (2019).

In a typical 3D reconstruction of an inspected volume, this one idivideded into voxels, that is, volumetric units with unknown uniform density. The voxel size depends on spatial resolution and on the amount of collected data. The measurement information can be subdivided in a number of directions (rays) which, in case the detector can be considered point-like, correspond to a solid angle. Given N_I rays and N_J voxels, muons in the ith ray cross an opacity X^i and Eq. (7) can be re-written as the following system of equations:

$$X^i = \sum_{j=1}^{N_I} L^{ij}\rho^j, i = 1, 2, \ldots, N_I \qquad (7')$$

where ρ^j is the density of the jth voxel and L^{ij} is the matrix of the length crossed by the ray i in the voxel j. In the 2D case, the problem was to determine the expected flux (transmission) and hence the mean opacity X^i for each direction i, given the measurement. Here, the inverse problem has the goal to determine the density values ρ^j ($j = 1, \ldots N_J$) that solve the system of Eq. (7') given the measured opacities $X^i(i = 1, \ldots N_I)$. The solution can be obtained by minimization procedures or iterative methods.

It has to be mentioned that some works[18,19] have shown that, when the detector size in not neligible with respect to the studied object, it is possible to obtain 3D information even with measurements from a single position.

In conclusion, this technique can provide both a radiographic (2D) and a tomographic (3D) description of very large, not accessible volumes.

Details on applications of this technique as well as the description of possible reconstruction algorithms are discussed in the following chapters.

3.2. *Absorption technique*

This technique is also based on energy losses and to the consequent absorption of low momentum muons in the target volume. However, in this case, the object to study is placed between two detectors, one supposed to have good tracking capability to determine quite precisely the direction of muons before they enter in the volume, the other supposed to act as a veto to identify particles that have been absorbed inside the target. Clearly, the object cannot be as big as the ones suitable to be studied with transmission technique since it would be impractical to place huge detectors on both sides of the target. On the other hand, the expected fraction of absorbed muons cannot be as small as in the case of thin and light objects. The technique is then suitable for studying middle-sized or very dense objects and can profit from large area detectors to increase the angular acceptance. Depending on the setup, the two detectors could be different (e.g., one detector is above the target and hence acts as a tracker for the incoming muons and the second is below and hence it acts just as a veto as shown in Fig. 5) or equal (e.g., to profit from muons coming from lateral directions).

It must be noticed that the cosmic rays have a minor component constituted by low-energy electrons that leave a signal on the 1st detector and that are immediately absorbed in the target. Therefore, they modify the absorption rate expected for muons, producing a distortion in the estimation of material density. However, there are methods to reduce their number and the related effect, as by inserting absorbers before the detectors.

As discussed above, this technique cannot be applied to geological or big archeological structures but, when applicable, it presents several advantages compared to the transmission technique. First, if the detector has a large or similar size with respect to the target, the directions of incoming muons can provide an intrinsic 3D reconstruction of the volume of interest. Secondly, the detector distance from the target can be minimal and hence the track extrapolation inside the volume to study can be quite precise and ensure a good spatial resolution. Furthermore, the incoming muon

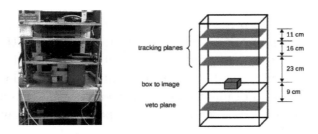

Figure 5. Example of a detector and its schematic drawing to test the absorption technique. The upper detector consists of three planes to ensure a good tracking while just a veto plane is used below the studied object. This figure is from Ref. 20.

flux (to be compared with free sky flux previously defined) and the transmitted one are measured simultaneously with a clear possibility to reduce systematic effects. In addition, if the two detectors are equal and both can provide tracking information, this technique could be combined with the scattering one with mutual advantages.

The formalism to reconstruct a density map from absorption measurements[20, 21] is similar to the one previously described in connection with the transmission technique: data are subdivided into rays or tubes (Lines of Response) and the volume in voxels, each corresponding to an unknown density. The basic information provided by the measures is the number of incoming and absorbed muons for each ray and one must solve a system of equations as the one reported in Eq. (7) or a similar one. Also in this case, the voxel size depends on the experimental resolution and on the dataset.

An example of a tomographic reconstruction of a lead structure with the detector shown in Fig. 5 is shown in Fig. 6. The lead structure is composed by lead bricks of $5 \times 5 \times 5$ or of $5 \times 5 \times 10\,\mathrm{cm}^3$ with two layers composed as in the figure. The reconstructed image reproduces roughly both the structure and the lead density.

Other good results with the absorption technique have been obtained on simulated data for an application of muography to spent nuclear fuel cask.[22] More details are given in Chapter 10.

In conclusion, this technique can provide a tomographic (3D) description of large or heavy targets.

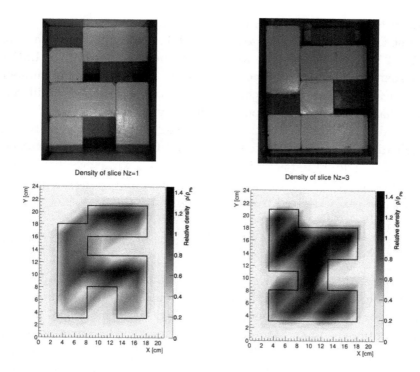

Figure 6. Example of reconstruction with absorption technique. In the upper pictures are shown the two layers composing the lead structure. In the lower panels the corresponding slicing of the tomographic reconstruction are shown and the lead boundaries are indicated by the black line. This figure is from Ref. 20.

3.3. *Multiple Coulomb Scattering (MCS) technique*

This technique follows an alternative approach being based on a different physical effect and it is suitable to study volumes similar to the ones of the absorption technique. The muon scattering technique has been proposed in 2003 in Los Alamos[23] and it exploits the dependency of the scattering angle distribution on the characteristic of the crossed material as expressed in Eq. (4) or in the simplified version

$$\sigma_\theta \approx \frac{b}{pc}\sqrt{l\lambda} \tag{4'}$$

where $b = 13.6\,\text{MeV/c}$, l is the crossed thickness, and $\lambda = \rho/X_0$ is defined as *linear scattering density* (LSD). In a rough approximation $(A \sim 2Z)$ valid within 20% for many elements, from Eq. (5) LSD is approximately the product of the density times the atomic number: $\lambda \sim \rho Z$ and hence, this technique can provide a map of a different material property with respect to the previous ones.

As discussed in Section 2.1 and reported in Eq. (6), the scattering angle distribution for cosmic muons is not Gaussian and the momentum of individual particles is in general unknown. Nevertheless, the variance of the distribution for cosmic muons that crossed a quantity $l\lambda$ can be easily derived from Eq. (6) and it is

$$\langle \theta^2 \rangle \propto l\lambda \left\langle \frac{1}{p^2} \right\rangle \approx l\rho Z \left\langle \frac{1}{p^2} \right\rangle \tag{9}$$

Therefore, the individual momentum can be substituted by a constant, the average value of the $1/p^2$ distribution of cosmic muons, and an approximate relation between measured scattering angles and the product of density and atomic number remains. This correspondence is valid for an element, but it holds also for compounds by considering their average atomic number. It must be taken into account that the absence of individual momentum measurement causes a loss of sensitivity and consequently for some specific application it may be necessary to measure or at least to provide some information about individual momenta.

MCS technique requires to measure the scattering angles of cosmic muons and hence it is necessary to provide at least two trackers to measure position and direction when muons enter and leave the target as sketched in Fig. 7.

Equation (9) foresees a proportionality between the variance of the scattering angle distribution projected onto a plane and the LSD of crossed materials given the same linear thickness l. Clearly, one should consider also the detector angular resolution σ_{det} which affects directly the measured variance $\langle \theta^2_{\text{meas}} \rangle$, but which can be taken into account as

$$\langle \theta^2_{\text{meas}} \rangle = \langle \theta^2 \rangle + 2\sigma^2_{\text{det}}.$$

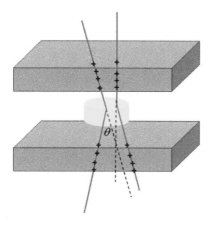

Figure 7.   Schematic example of MCS technique with two tracker detectors (gray boxes) measuring muon position and direction before and after the target (yellow disk).

However, there are effects that tend to weaken the linear dependency, in particular detector acceptance and energy spectrum distortion.[24] The former is related to the finite dimension of the detectors and, consequently, to the maximum scattering angle they can detect. It has been evaluated by simulating a set of cosmic muons crossing 10 cm of iron or alternatively of tungsten as shown in Fig. 8. Assuming a maximum scattering angle of 0.5 rad, the bias on the variance is 5% for iron but it grows to 15% for tungsten. The latter effect is related to the absorption of low energetic muons due to energy loss in the target and consequently to the distortion of cosmic muon spectrum. In particular, the average value of $1/p^2$ distribution considered in Eq. (9) changes and the variation depends on the crossed opacity. This effect has been measured and it is significant also for objects of the order of 10 cm as it can be seen in Fig. 9 where six cubic blocks of different materials, from aluminum to tungsten, are shown in the left panel. All blocks are 10 cm high (except tungsten, which is 8 cm high) and they have been inserted in the experimental apparatus described in Pesente (2009) and Checchia (2018)[24, 25] and in the appendix. A saturation in the reconstructed scattering densities is clearly visible and it is reproduced by a full

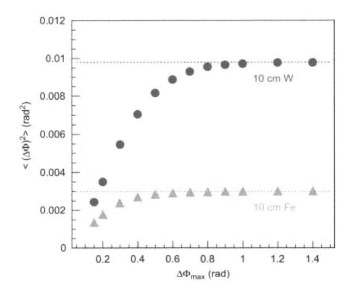

Figure 8. Effect on the variance of a scattering angle $\Delta\Phi$ of the maximum detectable angle $\Delta\Phi$max. The effect is shown for 10 cm of iron and for 10 cm of tungsten crossed by cosmic muons. This figure is from Ref. 24.

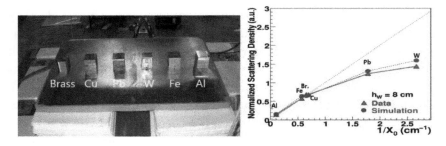

Figure 9. Picture of six blocks of different materials. Five blocks are 10 cm high, while the tungsten block is 8 cm high (left). The scattering density measured according to Eq. (9) (right). The scattering densities are normalized to give the best agreement for Fe, brass, and Cu. A saturation effect is evident. This figure is from Ref. 15.

simulation. The saturation is indeed due to the variation of the effective value of $\langle 1/p^2 \rangle$ since on simulation it is possible to remove the momentum dependency of scattering angles by plotting the quantity $\langle (p\theta)^2 \rangle$ which shows a linear dependency on target LSD for the same thickness. Considering that it is very difficult to provide a system capable of evaluating individual muon momenta, this result highlights how difficult is the evaluation of the absolute LSD value since it depends on the $\langle 1/p^2 \rangle$ value. A practical way to obtain LSD measurements with a reasonable precision is to use a known material to provide a proper calibration as discussed in Åström (2016).[26]

Since in Eq. (9) the measured angles are projected onto a plane, it is possible to provide measurements on the orthogonal planes which are almost independent. Consequently, a double measurement of $\theta_x \theta_y$ corresponds approximately to double the dataset or to reduce by a factor two the data-taking time. On the other hand, if for some technical reason the detector can provide a much better measurement of the scattering angle in one plane, the use of a single projection can be compensated by a more extensive data collection provided the tracking can ensure a measurement on the orthogonal coordinate sufficient to determine the particle path.

Despite difficulties related to saturation, if the trackers are larger or at least comparable with the target and they can measure both coordinates and directions projected onto orthogonal planes, MCS is a powerful technique to provide 3D images of an inaccessible volume.

A simple method for image reconstruction is based on the assumption that the scattering of individual muons is concentrated in a single point corresponding to the point of closest approach (POCA) of the tracks measured by the two detectors. One can reconstruct 3D images of objects by assigning to the POCA obtained for any muon a weight proportional to the square of the scattering projected angles, as can be seen in Fig. 10. However, this method tends to fail when the assumption is not realistic as in presence of several scattering centers along the muon trajectory as illustrated in Fig. 11.

An alternative method, based on maximum likelihood expectation maximization (MLEM), has been first proposed in Schultz (2003).[28] As the algorithms described for the previous techniques, MLEM divides the volume in N voxels having uniform LSD $\lambda_j (j = 1, \ldots, N)$. The input data, obtained from the collection of M muons, consist of the entry points and directions of particles on the volume to inspect as well as the measured projected angle $\Delta\theta$ and the displacement

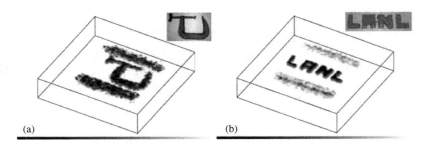

Figure 10. Images obtained with MCS data applying the POCA algorithm: a steel c-clamp (a) and the acronym "LANL" realized with one-inch lead stock. Steel beams used to support a plastic object platform are also visible in the reconstruction. This figure is from Ref. 27.

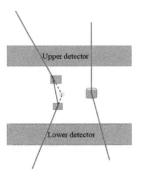

Figure 11. Schematic example of wrong reconstruction with POCA algorithm (yellow circle) in presence of two scattering centers (blue boxes).

Δx. This is schematically depicted in Fig. 12 where the muon path is supposed to be planar and perpendicular to the detectors. The measured displacement is included in the likelihood function since it brings information about the average vertical position of the scattering process and hence it improves the quality of the reconstruction. It is also possible to use the measured values (angle and displacement) in the orthogonal plane if compatible with the experimental setup.

The formalism is described in Schultz (2007),[29] Benettoni (2013),[30] and Bonomi (2020)[15] and here only a few points are recalled. (i) The scattering distribution is assumed to be Gaussian

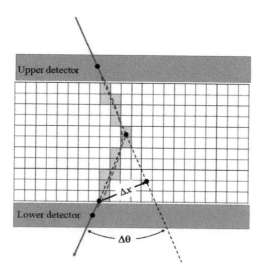

Figure 12. Schematic example of input data from a single muon to the MLEM algorithm. The dots on the detectors represent the position measurements, the red arrows the direction projected on the figure plane. The measured projected scattering angle and the displacement are shown as well. The colored voxels are the ones supposed to be crossed by the particle path.

even in absence of individual momentum measurement, and (ii) The muon path is unknown and hence a trajectory has to be assumed. A reasonable hypothesis is to follow straight lines crossing in the POCA as depicted by colored voxels in Fig. 12. Checks with detailed simulation show that in most cases it is a reasonable approximation. (iii) Limiting to the terms that include only angle measurement (hence neglecting for simplicity the displacement contribution) and considering that the ith muon has crossed n_i voxels, one obtains (excluding terms not containing λ_j)

$$\ln L \propto -\sum_{i}^{M} \left[\ln(C_i) + \frac{\theta^2}{C_i}\right] + \cdots \qquad (10)$$

where

$$C_i = 2\sigma_{\text{det}}^2 + \frac{b^2}{p_i^2} \sum_{j}^{n_i} l_{ij}\lambda_j \qquad (11)$$

and l_{ij} is the path of the ith muon inside the voxel j. The procedure is quite complex since the maximization of $\ln L$ should determine the value of N variables $\{\lambda\}$, which can be easily of the order of 10^6 if the volume to study is of the order of several cubic meters with a voxel size of a few cubic centimeters. In addition, the number of muons M can be several orders of magnitude larger than N. Therefore, the maximization must be obtained with an iterative process which can be quite slow[26] and require relevant computer resources. (iv) Since the individual muon momentum is in general unknown, Eq. (11) must be substituted by

$$C_i = 2\sigma_{\text{det}}^2 + b^2 \left\langle \frac{1}{p^2} \right\rangle \sum_j^{n_i} l_{ij}\lambda_j \qquad (11')$$

where the effective value of $\langle 1/p^2 \rangle$ has been introduced. This reduces the sensitivity to LSD variations and induces the effects of non-linearity and saturation described above.

Nevertheless, despite all difficulties and complexities, MLEM is a powerful tool to provide 3D images of a volume with several objects of different LSD as shown in the appendix.

Several other algorithms, often based on simulated data, to provide a tomographic reconstruction with MCS technique have been proposed. Their description can be found in the following incomplete reference list: Stapleton (2014),[31] Perry (2014),[32] Bandieramonte (2015),[33] Poulson (2017),[34] Chatzidakis (2018),[35] and Frazão (2019).[36] A combination of absorption and MCS techniques is also possible as reported in Morris et al. (2012).[37]

In conclusion, MCS is a technique that can provide a tomographic (3D) description of large or heavy targets with many practical applications as described in Chapter 10.

3.4. Muon-induced fission neutrons

When negative muons are absorbed in high-Z materials, they can be captured by the nuclear Coulomb field, reach the atomic ground state, and eventually interact with a proton in the nucleus producing a neutron and a neutrino. In fissile materials, the neutron can

Figure 13. Picture of the Los Alamos mini muon tracker (MMT) which consists of two trackers ("supermodules") based on 12 planes of drift tubes with a ∼60 cm gap. The LEU cube and the EJ-301 liquid scintillator detectors have been inserted on a wooden platform between the two supermodules. This figure is from Ref. 38.

induce a fission process producing in turn other neutrons, gamma rays, and fission fragments. The amount of neutrons depends on several geometrical factors and from the type and the quantity of fissile material. By correlating these neutrons with cosmic rays pointing to the fissile material through a coincidence window, one can produce a tagged image of the target. A proof-of-principle of this technique has been obtained in Los Alamos[38,39] with a 10 cm size cube of low enriched uranium (LEU: 19.7% ^{235}U). The cube and two neutron detectors have been inserted in the so-called mini muon tracker (MMT) that consists of an upper and a lower tracker (∼1.2 m × ∼1.2 m) with a 60 cm gap as shown in Fig. 13. The neutron detectors are two circular EJ-301 liquid scintillator detectors with a 12.5 cm diameter placed at 12.5 cm from the side of the cube. The scintillator signal shows a peak of hits between 350 ns and 650 ns after the muon time with prompt and delayed components as shown in Fig 14. The latter component has been fitted with an exponential and it is compatible with the lifetime of muonic atoms in uranium[40] plus the time scale of neutron multiplication (∼ 80 ns).

After a muon collection time from 1 to 44 hours, the image of the LEU cube could be reconstructed with all the muon techniques as

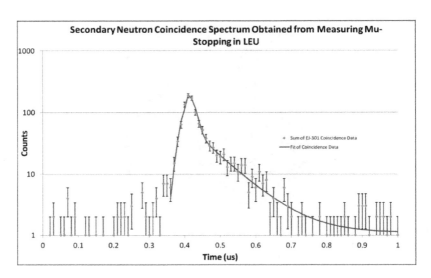

Figure 14. Time evolution of the neutron signal in coincidence with stopped muons. The exponential time constant is compatible with the lifetime of muonic atoms in uranium. This figure is from Ref. 38.

shown in Fig. 15. The MCS technique is clearly the most powerful for such small objects, but the presence of the LEU cube can be detected with the neutron-tagged technique in just 1 hour.

In conclusion, the feasibility of this additional technique can be very important for specific applications requiring the detection of fissile materials.

3.5. *Final remarks*

As discussed in this section, the different techniques can be used in different environments with different detectors and reconstruction algorithms. A summary of the various options is given in Table 2.

In addition to the techniques described above, the muon penetration properties can be exploited to align and monitor the stability of the relative position of different detectors. This practice, based on the track reconstruction capability of muon detectors

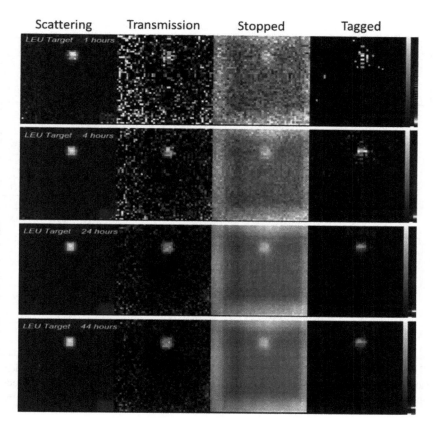

Figure 15. Images of 1,000 cm³ LEU with four cosmic muon techniques for different data collection times. This figure is from Ref. 39.

(see Chapters 4 and 5) has been used since decades in Particle Physics experiments. Since muons can easily penetrate the interposed material between detectors, one can use the measures to monitor the relative stability of the places where detectors are located even when a visual inspection is not possible. A description of possible applications of this technique, also known as muon metrology, can be found in Chapter 10.

Table 2. Summary comparison of various techniques.

Technique	Transmission	Absorption	MCS	n-tagged
Target size (largest dimension)	$>1\,\mathrm{km}$	$<20\,\mathrm{m}$	$<20\,\mathrm{m}$	$\sim\mathrm{m}$
Detector: minimum number	1	$2(1+\mathrm{veto})$	2	3
Detector: position w.r.t. target	Downstream	Upstream and Downstr.	Upstream and Downstr.	Upstream Downstr. Proximity n
Detector: minimum size (m^2)	1	$\geq$target	$\geq$target	$\geq$target (μ) $0.01-1(\mathrm{n})$
Image	No (rate) 2D (radiography) 3D (tomography)	3D	3D	3D
Minimum data taking time	Hours (rate) days	Hours	Minutes	Hours

Note: The term "rate" in the transmission column is referred to simple applications based on counting global muon rate. All values are indicative as they depend on specific applications.

4. Summary

Cosmic muography constitutes a bright example of how scientific knowledge and tools and instruments developed for fundamental research, namely Nuclear and Particle Physics, can find interdisciplinary and civil application. It is worth recalling the advantages in the use of these technologies: they exploit a natural, free, and continuous source without radiological hazards, they profit from well-known and reliable detectors, they can constitute the unique available technology when the thickness of the object to investigate does not allow the use of any other probe. Clearly, the main limitation is due to the muon flux, which is not compatible with fast or very fast measurements (minutes or seconds) and hence with rapidly moving objects. Nevertheless, an appropriate use in applications to study static or stationary situations, where a long time to collect data is possible, makes these technologies a good opportunity to produce positive impacts on our society.

Appendix: Examples of Tomographic Imaging with MSC Technique and MLEM Reconstruction

Some examples of images obtained with the INFN muon tomography demonstrator with MCS technique and the MLEM algorithm are presented and discussed. The demonstrator[24, 25] is based on two CMS barrel muon chambers[41] placed at a distance of about 160 cm and it has the capability to inspect a volume of more than 11 m^3. A picture of the system including chambers and various objects inserted in the inspection volume is shown in Fig. 16.

It is not necessary to describe all details of the two chambers, but it is relevant to mention that the angular resolution is much better ($\sim$2 $-$ 5 mrad) in one vertical plane (approximately the plane of the figure) than in the orthogonal one (>10 mrad). For

Figure 16. Experimental setup of the INFN muon tomography demonstrator. (1) Muon chambers, (2) iron I-beam structure, (3) iron slabs for calibration, (4) iron-plywood slabs for calibration, (5) aluminum tubes, and (6) concrete blocks. The red box indicates roughly the image reconstruction window.

this reason, only scattering angles and displacements measured in that plane are considered in the Likelihood function to maximize (described approximately in Eq. (10)). Nevertheless, the measures in the orthogonal plane are sufficient to provide a good particle path and to permit to split the volume in small size voxels.

For the present example, the image reconstruction regards a volume of 225 cm × 201 cm (on the horizontal plane) × 183 cm on the vertical direction for a total of 8.27 m³. The voxel size is 1.5 cm for a total of more than 2.4 million voxels. The data collection time was about 50 hours corresponding to more than 60 million events. The MLEM algorithm was run with 5,000 iterations. In Fig. 16, a red box indicates roughly the limits of the reconstruction area on the picture plane and several objects which enter in the inspection volume are numbered on the basis of their characteristics. The 2D reconstruction image of the horizontal plane is shown in Fig. 17 where all the objects visible in Fig. 16 are reconstructed with the addition of hidden details. The image is obtained by summing all voxels along the vertical coordinate and hence it is not possible to disentangle superimposed objects. A different view of the system is presented in Fig. 18 (top) where, together with the visible objects, are indicated the horizontal slices at different heights reproduced in panels (a)–(f). In this experimental setup, cosmic muons are

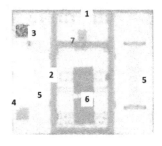

Figure 17. 2D reconstruction image of the horizontal plane. The objects listed in Fig. 16 are all reconstructed with the addition of (1) aluminum pad structures on the chamber and (7) metallic winch and chain.

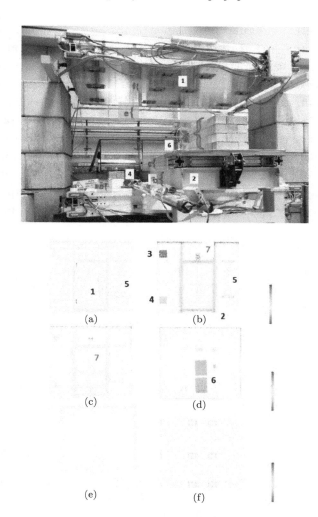

Figure 18. Picture with an alternative view of the experimental setup (top). The aluminum pad structures on the upper chamber plate are visible. The horizontal slices at different heights reproduced in panels (a)–(f) are also indicated. In panel (a) a shadow of the iron I-beam structure is visible, but the aluminum pads present in the lower chamber are well reconstructed, (b) all objects located in that vertical range are reproduced including the iron chain used to move, (c) the trolley made by aluminum supports and the light honeycomb structure where are placed, (d) the concrete blocks, (e) empty space, and (f) aluminum pads on the upper chamber. All colors are proportional to voxel LSD (in arbitrary units).

collected with an angular acceptance of at most ±45° from the zenith and hence the image resolution on the vertical direction is much worse than in the horizontal plane. Nevertheless, it is possible to disentangle overlapped objects including light aluminum structures on the chamber planes (panels (a) and (f)) and the trolley made by aluminum supports and a light honeycomb structure (panel (c)). The unique exception is due to the *shadow* of the heavy iron I-beam structure in panels (a) and (c), which is confused with the aluminum objects that are in contact with it. The images of calibration iron slabs in panel (b) and of concrete blocks in panel (d) are quite clear, while the empty space is almost transparent as shown in panel (e).

Acknowledgments

The author would like to thank P. Andreetto, E. Conti, and F. Gonella, for the contribution in producing the results shown in the appendix.

References

1. D.E. Groom, N.V. Mokhov, and S.I. Striganov, Muon stopping power and range Tables 10 MeV-100 TeV, *Atomic Data and Nuclear Data Tables*, **78**, 183–356, (2001). https://pdg.lbl.gov/2020/AtomicNuclear Properties/.
2. V. Kudryavtsev, Muon simulation codes MUSIC and MUSUN for underground physics, in: *Computer Physics Communications*, 180.3, pp. 339–346 (2009).
3. P.A. Zyla, *et al.*, (Particle Data Group) The review of particle physics, *Progess of Theoretical and Experimental Physics*, **2020**, 083C01, (2020). doi:10.1093/ptep/ptaa104.
4. G. Baccani, Development, testing and application to real case studies of a three-dimensional tomographic technique based on muon transmission radiography. Dipartimento di Fisica Università di Firenze PHC Thesis (2021). http://hdl.handle.net/2158/1239075
5. S. Agostinelli, *et al.*, Geant4 — a simulation toolkit, *Nuclear Instruments and Methods in Physics Research A*, **506**, 250–303, (2003). https://doi.org/10.1016/S0168-9002(03)01368-8
6. D. Heck, T. Pierog, and J. Knapp, CORSIKA: An air shower simulation program, Astrophysics (2012). Source Code Libr. ascl–1202.

7. A. Fedynitch, R. Engel, T.K. Gaisser, F. Riehn, and T. Stanev, Calculation of conventional and prompt lepton fluxes at very high energy, in: *EPJ Web of Conferences, vol. 99, EDP Sciences*, p. 08001 (2015). https://doi.org/10.1051/epjconf/20159908001

8. C. Hagmann, D. Lange, and D. Wright, Cosmic-ray shower generator (CRY) for Monte Carlo transport codes, in: 2007 IEEE Nuclear Science Symposium Conference *Record, IEEE*, (2007). http://dx.doi.org/10.1109/nssmic.2007.4437209

9. D. Pagano, *et al.*, EcoMug: An efficient Cosmic muon generator for cosmic-ray muon applications, *Nuclear Instruments and Methods in Physics Research, A*, 1014165732, (2021).

10. P. Biallass, and T. Hebbeker, Parametrization of the cosmic muon flux for the generator CMSCGEN, *arXiv preprint* arXiv:0907.5514 (2009).

11. L. Bonechi, *et al.*, Development of the ADAMO detector: Test with cosmic rays at different zenith angles, in: *29th International Cosmic Ray Conference*, vol. 9, p. 283 (2005).

12. E.P. George, Cosmic rays measure overburden of tunnel, *Commonwealth Engineer*, **455**, (1955).

13. L.W. Alvarez, *et al.*, Search for hidden chambers in the pyramids, *Science*, **167**(3919), 832–839, (1970).

14. Saracino, *et al.*, Imaging of underground cavities with cosmic-ray muons from observations at Mt. Echia (Naples), *Scientific Reports*, **7**, 1181, (2017). doi:10.1038/s41598-017-01277-3.

15. G. Bonomi, P. Checchia, M. D'Errico, D. Pagano, and G. Saracino, Applications of cosmic-ray muons, *Progress in Particle and Nuclear Physics*, **112**, 103768, (2020). https://doi.org/10.1016/j.ppnp.2020.103768

16. H.K.M. Tanaka, H. Taira, T. Uchida, M. Tanaka, M. Takeo, T. Ohminato, Y. Aoki, R. Nishitama, D. Shoji, and H. Tsuiji, Three-dimensional computational axial tomography scan of a volcano with cosmic ray muon radiography, *Journal of Geophysics Research*, **115**, B12332, (2010). doi:10.1029/2010JB007677.

17. K. Morishima, M. Kuno, A. Nishio, *et al.*, Discovery of a big void in Khufu's Pyramid by observation of cosmic-ray muons, *Nature*, **552**, 386–390, (2017). https://doi.org/10.1038/nature24647

18. L. Bonechi, R. D'Alessandro, N. Mori, and L. Viliani, A projective reconstruction method of underground or hidden structures using atmospheric muon absorption data, *JINST*, **10**, P02003, (2015). https://doi.org/10.1088/1748-0221/10/02/p02003

19. L. Bonechi, R. D'Alessandro, and A. Giammanco, Atmospheric muons as an imaging tool, *Reviews in Physics*, **5**, 100038, (2020). https://doi.org/10.1016/j.revip.2020.100038

20. G. Baccani, Muon absorption tomography of a lead structure through the use of iterative algorithms, *JINST*, **15**, P12024, (2020).
21. S. Vanini, *et al.*, Muography of different structures using muon scattering and absorption algorithms, *Philosophical Transactions of the Royal Society A*, **377**, 20180051, (2018). https://doi.org/10.1098/rsta.2018.0051
22. D. Ancius, *et al.*, ESARDA 41st Annual Meeting, Symposium on Safeguards and Nuclear Material Management, 14–16 May, Stresa (Italy) 142 (2019). doi:10.2760/159550.
23. K.R. Borozdin, *et al.*, Radiographic imaging with cosmic-ray muons, *Nature*, **422**, 277, (2003).
24. S. Pesente, *et al.*, First results on material identification and imaging with a large-volume muon tomography prototype, *Nuclear Instruments and Methods in Physics Research A*, **604**, 738–746, (2009). doi:10.1016/j.nima.2009.03.017
25. P. Checchia, *et al.*, INFN muon tomography demonstrator: Past and recent results with an eye to near-future activities, *Philosophical Transactions of the Royal Society*, **377**, 20180065, (2018). http://dx.doi.org/10.1098/rsta.2018.0065
26. E. Åström, *et al.*, Precision measurements of Linear Scattering Density using Muon Tomography, *JINST*, **11**, P0701, (2016).
27. L.J. Schultz, *et al.*, Image reconstruction and material Z discrimination via cosmic ray muon radiography, *Nuclear Instruments and Methods in Physics Research A*, **519**, 687–694, (2004).
28. L.J. Schultz, *Cosmic Ray Muon Tomography*, Ph.D. dissertation (Portland, OR: Portland State University, 2003).
29. L.J. Schultz, *et al.*, Statistical reconstruction for cosmic ray muon tomography, *IEEE Transactions on Image Processing*, **16**, 8, (2007).
30. M. Benettoni, *et al.*, Noise reduction in muon tomography for detecting high density objects, *JINST*, **8**, P12007, (2013). doi:10.1088/1748-0221/8/12/P12007.
31. M. Stapleton, *et al.*, Angle statistics reconstruction: A robust reconstruction algorithm for Muon Scattering Tomography, *JINST*, **9**, P11019, (2014). doi:10.1088/1748-0221/9/11/P11019.
32. J.O. Perry, *et al.*, Analysis of the multigroup model for muon tomography based threat detection, *Journal of Applied Physics*, **115**, 064904, Detection *Journal of Applied Physics* 115, 064904, (2014). doi:10.1063/1.4865169.
33. M. Bandieramonte, *et al.*, Clustering analysis for muon tomography data elaboration in the Muon Portal project, *Journal of Physics: Conference Series*, **608**, 012046, (2015).

34. D. Poulson, *et al.*, Cosmic ray muon computed tomography of spent nuclear fuel in dry storage casks, *Nuclear Instruments and Methods in Physics Research A*, **842**, 48–53, (2017). http://dx.doi.org/10.1016/j.nima.2016.10.040

35. S. Chatzidakis, *et al.*, A generalized muon trajectory estimation algorithm with energy loss for application to muon tomography, *Journal of Applied Physics*, **123**, 124903, (2018). https://doi.org/10.1063/1.5024671

36. L. Frazão, *et al.*, High-resolution imaging of nuclear waste containers with Muon Scattering Tomography, *JINST*, **14**, P08005, (2015). https://doi.org/10.1088/1748-0221/14/08/P08005

37. C.L. Morris, *et al.*, Obtaining material identification with cosmic ray radiography, *AIP Advances*, **2**, 042128, (2012). https://doi.org/10.1063/1.4766179

38. J. Bacon, *et al.*, Muon Induced Fission Neutrons in Coincidence with Muon Tomography *LA-UR*-13-28292 (2013).

39. E. Guardincerri, *et al.*, Detecting special nuclear material using muon-induced neutron emission, *Nuclear Instruments and Methods in Physics Research A*, **789**, 109–113, (2015). http://dx.doi.org/10.1016/j.nima.2015.03.070

40. H. Hänscheid, P. David, J. Konijn, *et al.*, Muon capture rates in ^{233}U,^{234}U,^{235}U,^{236}U,^{238}U, and ^{237}Np. *Z. Physik A — Atomic Nuclei*, **335**, 1–8, (1990). https://doi.org/10.1007/BF01289340

41. S. Chatrchyan, The CMS Collaboration *et al.*, The CMS experiment at the CERN LHC, *JINST*, **3**, S08004, (2008).

42. V. Tioukov, I. Kreslo, Y. Petukhov, and G. Sirri, The FEDRA — Framework for emulsion data reconstruction and analysis in the OPERA experiment, *Nuclear Instruments and Methods A*, **559**, 103–105, (2006).

43. L. Cimmino, *et al.*, *Sci. Rep.* **9**, 2974 (2019).

Chapter 4

Emulsion Detectors for Muography

Akira Nishio

*Institute of Materials and Systems for Sustainability,
Nagoya University, Furo-cho, Chikusa-ku,
Nagoya, 464-8601, Japan
nishio@flab.phys.nagoya-u.ac.jp*

Emulsion detectors are one of the most suitable detectors for muon radiography. The detectors are compact and do not require a power supply, making it suitable for measurements in narrow space or outdoors where there is no power supply. In this chapter, we will introduce the basics of emulsion detectors and recent updates to emulsion detectors more suitable for muon radiography.

1. Introduction

Nuclear emulsion is a film-type three-dimensional (3D) particle track detector that uses silver halide photosensitive material technology and has a long history of use in particle physics experiments.[1] The nuclear emulsion detector is suitable as a detector for muon radiography because it is compact and does not require a power source. Nuclear emulsion detectors have been used for various muon radiography measurements, starting from the measurements of Showa-shinzan[2] and Mt. Asama volcanoes.[3] In this chapter, we introduce the basics of nuclear emulsion technology and the development of nuclear emulsions more suitable for muon radiography.

2. Emulsion Detectors

2.1. *History*

The history of nuclear emulsions as radiation detectors dates back to more than 100 years. In this subsection, I will review the achievements in particle physics that have been made using nuclear emulsion detectors, and then describe the technological developments in nuclear emulsion that have made these discoveries possible.

In 1896, Becquerel studied fluorescence from uranium ore. When he placed the uranium ore and a photographic plate together in his desk drawer, the plate turned black without being exposed to light, leading to the accidental discovery of radioactivity.[4] In 1910, Kinoshita discovered that when a photographic plate was irradiated with alpha rays, each alpha particle produced a different silver particle.[5] In 1911, Reinganum recorded for the first time the track of alpha particles on a photographic plate as a series of developed silver particles.[6] The photographic plate used as a radiation track detector was later called a nuclear emulsion.

Since then, nuclear emulsion detectors have produced many important results in particle physics experiments. In 1947, Powell *et al.* discovered the pion mesons predicted by Yukawa in cosmic rays using a modified nuclear emulsion.[7] In 1971, Niu and his colleagues discovered a charm particle in a nuclear emulsion on an aircraft.[8] Niu continued to use an accelerator to develop a combined nuclear emulsion and electronics detector experiment (Fermilab-E531, CERN-WA75), and contributed to the elucidation of the properties of charm particles.[9, 10]

Subsequently, accelerator-based neutrino experiments were developed by exploiting the high positional resolution of nuclear emulsions. In the CHORUS experiment, the detector was exposed to the neutrino beam for 4 years, from 1994 to 1997, while examining oscillations from muon neutrinos to tau neutrinos at a neutrino flight distance of 1 km.[11] The DONUT experiment in 1997, in which neutrino beams containing tau neutrinos were exposed to the detector, succeeded in capturing tau neutrinos in nuclear emulsion.[12, 13] And the OPERA experiment[14] was conducted from 2008 to 2012 to verify the oscillation from muon neutrinos to tau neutrinos by the appearance method at a neutrino flight distance of 730 km. By 2018, 10 tau neutrinos produced by neutrino oscillations were detected, confirming neutrino oscillations in the form of a tau neutrino appearance.[15]

Continuous improvements in nuclear emulsion technology have been essential for advances in particle physics. Nuclear emulsion originated in the technology of silver halide photosensitive materials and has been improved to make them suitable for detecting elementary particles.

At the beginning of the 20th century, Henri Becquerel and Kinoshita used ordinary photographic plates, which were not suitable for detecting particle tracks. Thus, only slow particles, such as alpha rays, were observed. In 1927, Mysssowsky and Tschiskow created a nuclear emulsion with a 50-μm thick emulsion layer to observe longer tracks.[16] In 1934, Blau and Wambacher reported that pinacryptol yellow increased proton sensitivity.[17] In 1947, Powell *et al.* first discovered a pion, but their plates could not capture electrons in e. In the following year, Kodak increased the sensitivity of the nuclear emulsion to capture the minimum ionized particles (MIP),[18] and Powell *et al.* succeeded in capturing all e decays using the new emulsion plates (Fig. 1).[19] Based on this nuclear emulsion, Kodak released the NT4 nuclear emulsion in 1948, and Ilford released the G5 nuclear emulsion in 1949, which were used as standards for nuclear emulsion experiments. The ET7A nuclear emulsion released

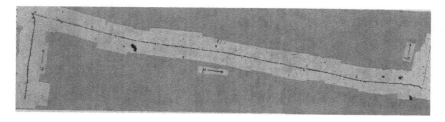

Figure 1. Decay of $\pi \to \mu \to$ e recorded on a nuclear emulsion. Redrawing Fig. 6 of Ref. 19.

by Fujifilm in 1957 had the same sensitivity to MIP as the NT4 and G5 nuclear emulsions.

Because cosmic ray experiments require large-area nuclear emulsion, the conventional method using only a thick (600 μm) emulsion layer (called a pellicle) was replaced by a nuclear emulsion with a thin emulsion layer of 50–70 μm on both sides of a transparent plastic film. The pellicle caused large distortions in the tracks, but by attaching the emulsion to the plastic, the emulsion was fixed, enabling accurate measurements. Using an emulsion cloud chamber (ECC) that consists of alternating layers of nuclear emulsion and scatterers, Niu *et al.* discovered charm particles in 1971.[8]

The OPERA experiment, which started in 2008, required 100,000 m^2 of nuclear emulsion targets, thus Nagoya University and Fujifilm collaborated to make the necessary improvements. At first, it was necessary to reduce the cost of the film. To achieve this goal, the volume occupancy of silver halide crystals was reduced to 30%. On the other hand, in order to ensure the grain density of silver particles along the track, improvements were made to increase the quantum sensitivity of the silver halide crystals. Secondly, the thickness that could be applied at one time was 21 μm by machine coating in an industrial plant of photographic film, thus the emulsion layer was secured at 42 μm by applying two coats. In addition, a protective layer of 1 μm consisting of gelatin was formed on the surface by the multilayer coating technology. At last, because the OPERA experiment required a low background nuclear emulsion, a method to efficiently eliminate the noisy tracks accumulated between the time of film production and the start of the experiment was developed. To eliminate the noise tracks, a "refreshing technique" was adopted, in which the nuclear emulsions were stored at high temperature and high humidity. A chemical called 5-methylbenzotriazole was found to enhance this refreshing effect and was added to the emulsion. The nuclear emulsion jointly developed by Nagoya University and Fujifilm is called OPERA film.[20, 21] Photo of OPERA detector using 100,000 m^2 OPERA films is shown in Fig. 2.

Due to the impact of digitalization starting from the beginning of the 21st century, photo manufacturers worldwide have reduced their

Figure 2. Photo of OPERA detector. 100,000 m² OPERA films were used.

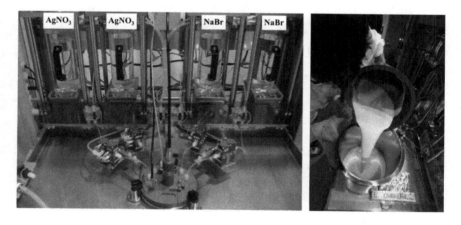

Figure 3. Emulsion gel facility at Nagoya University.

production and development of silver halide photography, hindering the continuous development and supply of nuclear emulsions by companies. In 2010, Nagoya University, with the cooperation of Fujifilm alumni, installed a nuclear emulsion manufacturing system at the university to enable the development and production of nuclear emulsions (Fig. 3). This enabled users to develop their own nuclear emulsions according to their applications.

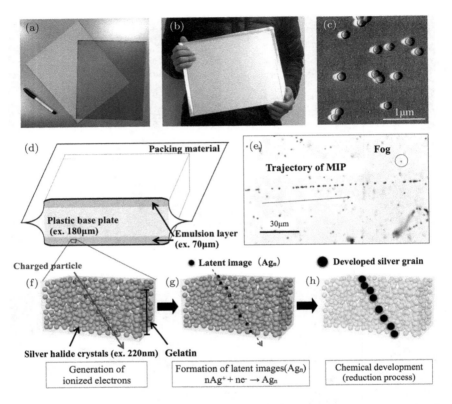

Figure 4. (a) Nuclear emulsion before (left) and after (right) chemical development. (b) Nuclear emulsion vacuum-packed with an aluminum-laminated envelope. (c) Micrograph (SEM) of silver-halide crystals. (d) Cross section of nuclear emulsion. (e) Optical micrograph of nuclear emulsion (dashes crossing the center of the figure represent a track of MIP and the circle surrounds a fog). (f) Enlarged image of the emulsion layer and generation of ionized electrons by charged particles. (g) Formation of a series of latent images that are the source of a track. (h) Enlarged silver grains from latent image by chemical development process. Redrawing Fig. 1 of Ref. 22.

2.2. Detection mechanism

The nuclear emulsion is a film-type detector. Figure 4(a) shows a nuclear emulsion before (left) and after (right) the development process. As the nuclear emulsion before development is sensitized by light, it is placed in an aluminum-laminated bag to shield it from light and to protect from the outside environment (water, dust, etc.). The film is vacuum-packed in an aluminum-laminated bag such that the shape of the film can be seen, as shown in Fig. 4(b). Figure 4(d) shows the cross-section of a packed nuclear emulsion. The emulsion layer is present on both sides of a transparent plastic base that allows light to pass through. The plastic base is made of a material with low birefringence, such as polystyrene. The thickness of the base is typically tens to hundreds of microns, with 180–500 μm being used in recent years. The emulsion layer consists of silver halide crystals tens to hundreds of nanometers in diameter dispersed by gelatin (Fig. 4(f)). Figure 4(c) shows a scanning electron microscopy (SEM) image of a silver halide crystal on a nuclear emulsion with a crystal diameter of approximately 220 nm. In the nuclear emulsion, each crystal functions as a sensor. The thickness of the emulsion layer ranges from tens to hundreds of microns, and 70 μm has been used in recent years.

The electrons in the valence band of the silver halide crystal are excited by ionization along the trajectory of the charged particle (Fig. 4(f)). The excited electrons associate with the interstitial silver ions in the crystal to form a group of silver atoms, called a latent image (Fig. 4(g)). As the latent images are too small to be observed, they are amplified by the development process to form silver particles of a size that can be observed with an optical microscope (Fig. 4(h)).

2.3. *Performance*

The nuclear emulsion is characterized by its high spatial resolution because it is a chemical detector using crystals. The measurements showed that the principle positional resolution is approximately 50 nm for a 200 nm crystal (Fig. 5).[23]

Therefore, if the thickness of the base is 200 μm, the angular resolution obtained in principle is 0.5 mrad. This is the principle resolution of the nuclear emulsion itself, and the actual resolution of the measurement system includes the image processing in the scanning system described in what follows.

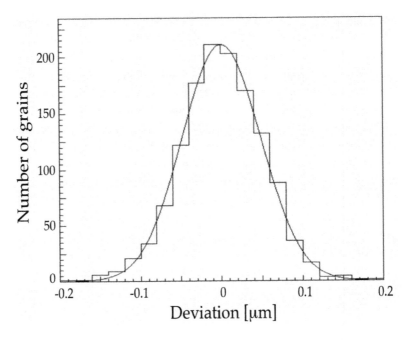

Figure 5. Positional resolution of nuclear emulsion. Deviation of grains from a linear-fit line for horizontal tracks. Redrawing Fig. 3 of Ref. 23.

2.4. *Scanning system*

Nuclear emulsion records an enormous amount of information because it records the trajectory with sub-micron positional accuracy. Until the 1970s, studies using nuclear emulsion required considerable time and effort for analysis because the data had to be manually interpreted. In the mid-1970s, Niwa *et al.* devised the idea of an automatic track selector,[24] and in 1981, Aoki *et al.* realized the first automatic track selector,[25] which was later put to practical use by Nakano.[26] Since then, the speed of the track scanning system has been continuously improved.[27–29] Figure 6 shows scanning system generation and scanning speed. The hyper-track selector (HTS)[30] can read tracks at a rate of $4,700 \, \mathrm{cm}^2/\mathrm{h}$. This means that one HTS can read and analyze approximately $1,000 \, \mathrm{m}^2$ of nuclear emulsion per year. The photo of HTS installed at Nagoya University is shown in Fig. 7.

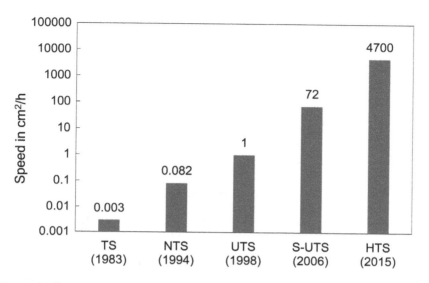

Figure 6.   Scanning system generation and scanning speed. Redrawing Fig. 2 of Ref. 30.

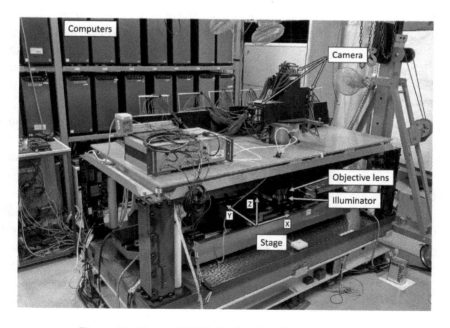

Figure 7.   Photo of HTS. Redrawing Fig. 3 of Ref. 30.

A. Nishio

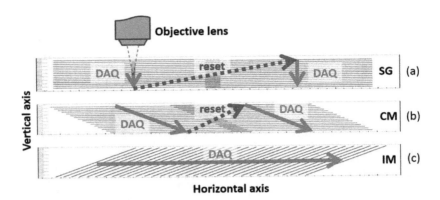

Figure 8. Ilustration of (a) Stop&Go (SG), (b) Continuous Motion (CM), and (c) the Inclined Motion (IM) scanning techniques. Redrawing Fig. 1 of Ref. 32.

In Europe, Armenise *et al.* (2005)[28] and Arrabito *et al.* (2006)[29] developed a scanning system called the European Scanning System (ESS) with a speed of $20\,\mathrm{cm}^2/\mathrm{h}$. Alexandrov *et al.* (2017, 2019)[31,32] also developed a new method of reconstructing stereoscopic images using an inclined focusing optical system with several commercial objectives to improve scanning speeds. By realizing a data taking with continuous motion technique, the scanning speed reached $190\,\mathrm{cm}^2/\mathrm{h}$. And they demonstrated the potential for scanning speeds in the thousands of cm^2/h with inclined motion technique (Fig. 8).

Figure 9 shows schematic representation of nuclear emulsion scanning. The information obtained by the scanning system is the position, angle, and density of silver grains of the track recorded on the nuclear emulsion. For automatic tracking recognition, a tomographic image is acquired by shifting the focal plane of the microscope in the emulsion layer. When the tomogram is horizontally shifted and the alignment of the developing silver particles is added together, only the angles at which the tracks are present show high values, and they are recognized as tracks if they exceed a certain threshold.

In the case of HTS, an angular accuracy of 2 mrad has been achieved for vertical tracks using a 200-nm crystal with a 180-μm base (Fig. 10). In addition, more than 98% detection efficiency was achieved under good conditions.[30]

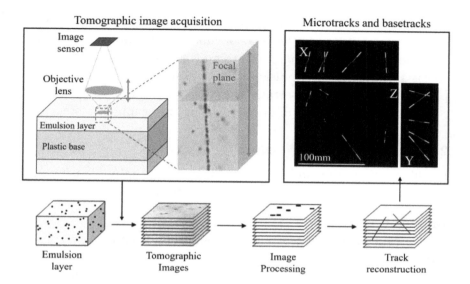

Figure 9. Schematic representation of nuclear emulsion scanning. Redrawing Fig. 7 of Ref. 27.

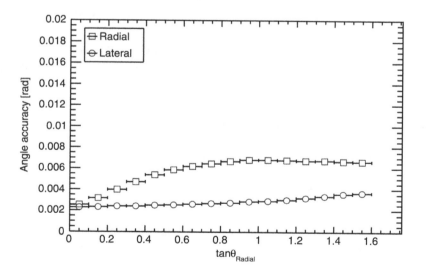

Figure 10. Angle dependence of angle accuracy for base tracks read by HTS. Redrawing Fig. 22 of Ref. 30.

Owing to the drastic improvement in scanning speed, the use of nuclear emulsion in the field of muon radiography has become possible. The nuclear emulsion technology for muon radiography consists of the nuclear emulsion itself, which is based on silver halide photography, automatic track selector, and track reconstruction technologies from image data obtained by the scanning system.

2.5. *Measurement procedure*

In this section, the procedure of muon radiography measurement using a nuclear emulsion is explained in detail.

Emulsion production
Manufacturing of the nuclear emulsion from raw materials.
Coating
Application of the nuclear emulsion to a plastic base.
Moisture conditioning and packing
Damp the nuclear emission under a constant humidity environment, and pack it in a light-shielding and moisture-proof aluminum laminate bag to control its water content.
Installation and surveying
Set the nuclear emulsion toward the object to be measured using an installation jig. The position and direction of the detector and the 3D outline of the measurement target are measured.
Measurement (accumulation of cosmic rays)
Accumulate cosmic rays. It is important that the detector does not move during this process.
Collect
Collect the detector.
Chemical Development
Chemically develop the nuclear emulsion using a developing solution.
Cleaning
Remove the silver deposited on the surface of the nuclear emulsion. In the case of surface-coated nuclear emulsions, this is not necessary because silver does not precipitate.

Swelling

As the silver halide dissolves and the thickness is reduced by half due to chemical development, the nuclear emulsion is swollen and thickened with water, and then immobilized using glycerin solution.

Scanning

Using a scanning system, read out the tracks recorded on the nuclear emulsion.

Track reconstruction

Select only the signal tracks from the read tracks.

Imaging and analysis

Use the angular distribution of cosmic ray muons, the 3D structure of the object, and simulations to visually and analytically show the density structure information of objects.

3. Development of Emulsion Detector for Muon Radiography

Because muon radiography is performed in an environment different from that of conventional particle experiments, it is necessary to develop a detector adapted to this environment. Specifically, it is necessary to have a detector that can stably operate in various environments, and a detector structure that can screen out particles other than muons that serve as background. In addition, mass production of the detectors and further acceleration of the readout system are required to manage a larger number of observation targets. In this section, we discuss how nuclear emulsions can meet these requirements.

3.1. *Emulsion stability*

In muon radiography, the observation targets are often located outdoors, and the nuclear emulsion is expected to be exposed to a variety of environments. This environment is completely different from a stable observation environment, such as an experimental hall in a particle experiment, where nuclear emulsion was used in the past. Originally, nuclear emulsions were suitable for outdoor observation because they do not require a power supply, are waterproof and dust-proof by being packed with packaging materials, and are lightweight. However, due to issues related to temperature resistance and long-term characteristics, the observation period, season, and location have been limited.

The issues of long-term observation using nuclear emulsion are latent image fading, fog increase, and sensitivity degradation. When a nuclear emulsion is stored for a long time before it is developed, the latent image becomes smaller due to oxidation ($Ag_n \rightarrow Ag_{n-1} + Ag^+ + e^-$). This is called latent image fading. As the latent image fading progresses, the grain density/100 μm (GD), which is the number of developed silver particles along the tracks, decrease. A fog is the result of the generation of a development starting point on the surface of the crystal due to a cause other than the track. The fog is the noise when observing the track and is evaluated using the fog density/1,000 μm^3 (FD). Sensitivity degradation refers to a decrease in the sensitivity of silver halide crystals to charged particles over time. As the sensitivity degradation progresses, the GD decreases, and the tracks become difficult to recognize.

Nishio et al.[22] investigated the factors that affect the long-term characteristics of nuclear emulsions to enable long-term observation and year-round outdoor use. They found that the important factors that affected long-term stability are gelatin, added chemicals, and packing materials. Figures 11 and 12 show the latent image fading characteristics, Figs. 13 and 14 show the fog characteristics, and Fig. 15 shows the sensitivity degradation characteristics of the new stable nuclear emulsion developed at Nagoya University. The new stable nuclear emulsion achieved the target performance of GD $\geq$ 30 and FD $\leq$ 5 at 30 for more than half a year.

The development of a new stable nuclear emulsion has enabled to perform continuous observations indoors for a year and outdoors during summer. In fact, muon radiography observations at the Khufu pyramids[33] in the Egyptian desert were performed using a stable nuclear emulsion.

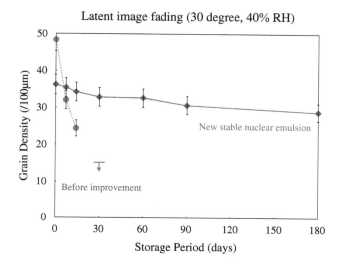

Figure 11. Latent image fading characteristics of nuclear emulsion at 30, 40% relative humidity (RH). In the samples before improvement, after day 30, no track could be recognized due to the low GD. Redrawing Fig. 11 of Ref. 22.

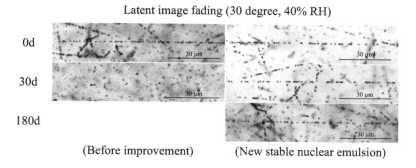

Figure 12. Optical micrographs of nuclear emulsion exposed to the electron beam (several tens of MeV) which were developed on days 0, 30, and 180. Latent image fading characteristics at 30, 40% RH. The length of the black bar is 30 μm. Redrawing Fig. 12 of Ref. 22.

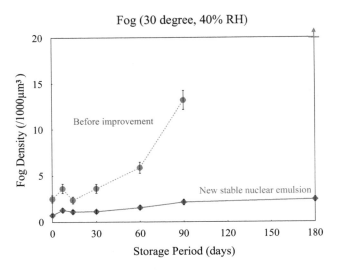

Figure 13. Fog characteristics of nuclear emulsion at 30, 40% RH. In the samples before improvement, on day 180, it was not possible to evaluate the FD due to the large amount of fog. Redrawing Fig. 14 of Ref. 22.

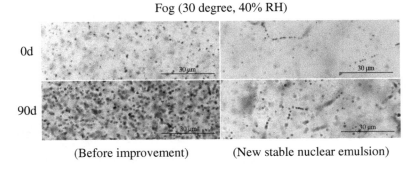

Figure 14. Optical micrographs of nuclear emulsion which were developed on days 0 and 90. Fog increase characteristics at 30, 40% RH. The length of the black bar is 30 μm. Redrawing Fig. 15 of Ref. 22.

3.2. Background rejection

3.2.1. Selection of tracks during observation

In nuclear emulsions, the accumulation of tracks starts immediately after production. Figure 16 shows the method of identifying tracks

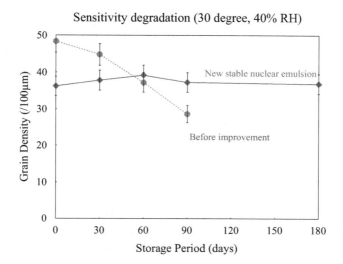

Figure 15. Sensitivity degradation characteristics of nuclear emulsion at 30, 40% RH. In the samples before improvement, on day 180, it was not possible to evaluate the GD due to the large amount of fog. Redrawing Fig. 17 of Ref. 22.

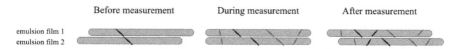

Figure 16. Method of identifying tracks during measurement. The tracks accumulated during the measurement are shown in red, and it is possible to distinguish them from the other tracks shown in black.

during measurement. Two nuclear emulsions are placed on top of each other during installation, and by considering the continuous tracks recorded on those two films, the tracks that are not in the installation can be discriminated. After the measurement, the two nuclear emulsions are separated again such that they can be distinguished from each other during the measurement.

3.2.2. *Removal of low-energy noise*

When the measurement target is a thick object such as a volcano, the number of transmitted muons decreases, and the influence of noise from cosmic ray components such as electrons and protons cannot be neglected. The noise can be removed by installing the underground detector, where these particles cannot penetrate.[34] In addition, protons which are particles with high ionization loss, and muons which are the minimum ionized particles, can be distinguished from each other by using the difference in GD in the nuclear emulsion.

In addition, a multilayer detector of a nuclear emulsion is effective in removing noise. By using an ECC that consists of alternating layers of nuclear emulsion and scatterers such as lead, momentum can be measured by determining the amount of multiple electromagnetic scattering from the position and angle at which charged particles are scattered before and after the scatterer.[35] By measuring the ionization loss and momentum of the charged particles detected by the nuclear emulsion, it is possible to discriminate the particle species and remove the low-energy noise component contained in cosmic rays, which is the noise in muon radiography observations. The accuracy of the momentum measurement by ECC is determined by the thickness of the thin heavy material and the number of layers. Nishiyama et al. (2016)[36] reported that the noise for the muon radiography of Showa-shinzan can be reduced by using OPERA-type ECC with 20 emulsion films and nine 1-mm-thick lead plates. This procedure can remove most of the tracks with less than 1 GeV/c (Fig. 17).

3.2.3. Effects of environmental radiation

Nuclear emulsion accumulates environmental radiation tracks caused by radiation nuclides in nature immediately after manufacture. The accumulated environmental radiation tracks become noise in the recognition of muon tracks, thus, it is desirable to minimize the accumulation of unnecessary tracks. If a long period of time elapses between the manufacture of the nuclear emulsion and its use for measurement, a large number of tracks will have accumulated. In such a case, it is desirable to expose the nuclear emulsion to a high-temperature and high-humidity environment for a certain period of time before use to intentionally accelerate the latent image fading and perform a refreshing process[20, 21] to eliminate the tracks. Figure 18 shows photomicrographs of nuclear emulsion with accumulated

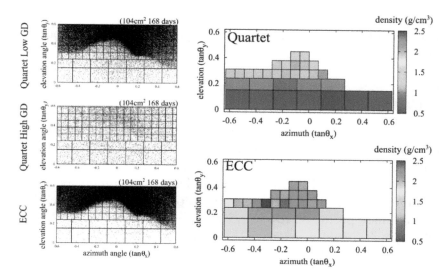

Figure 17. Measurement of background rejection with ECC at Showa-shinzan. Redrawing Fig. 10 and Fig. 11 of Ref. 36.

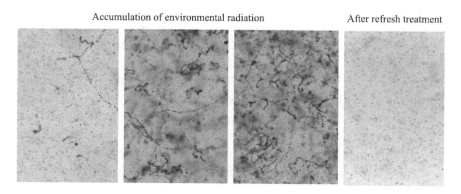

Figure 18. Photomicrographs of nuclear emulsion with accumulated environmental radiation, and elimination of tracks by refreshing.

environmental radiation, and elimination of tracks by refreshing. Moreover, during the observation, if the observation is performed for a long period of time in a place with high environmental radiation levels, the noise will increase and adversely affect the measurement. In such cases, it is necessary to protect the nuclear emulsion from low-energy environmental radiation by placing it in a shielding material or removing the noise during the post-development and analysis processes.

3.3. Mass production

Muon radiography measurements are expected to be used in a variety of applications in the future, measurements from multiple points, and measurements of huge structures with a thickness of more than 1,000 m, such as Mt. Fuji in Japan. These measurements require a large number of detectors, and in principle, nuclear emulsion can be mass-produced to meet future measurement needs. In fact, 100,000 m^2 of nuclear emulsion have been produced for the OPERA experiment using Fujifilm's production line.

As the demand for photographic films is decreasing and there is no guarantee of continuous cooperation from photographic manufacturers, it is important for users to have their own development and supply system for nuclear emulsion for the continuous development of research. To develop and produce nuclear emulsions, Nagoya University has been constructing emulsion manufacturing and mechanical coating equipment. Including the requirements of various physics experiments as well as muon radiography, the demand for nuclear emulsion of 1,000 m^2 per year is expected to increase in the future, and production facilities are being developed to meet the demand. In 2010, Nagoya University installed an experimental nuclear emulsion production system capable of producing approximately 0.5 kg of nuclear emulsion with the cooperation of Fujifilm alumni. As of 2021, a mass production machine that is 30 times larger than the first test machine is being set up at Nagoya University. It is capable of producing 15 kg of nuclear emulsion at a time, which is equivalent to approximately 10 m^2 of nuclear emulsion with 70-μm emulsion layers on both sides.

In addition, Nagoya University has a manual emulsion coating facility since around 2010 and has been providing nuclear emulsion for various experiments. The current manual application facility is capable of producing double-sided nuclear emulsions with a thickness

Figure 19. Machine coating system for nuclear emulsion under construction at Nagoya University.

of 70 μm at a maximum speed of 9 m²/week. To establish a supply system for nuclear emulsion of 1,000 m² per year, a machine coating system for nuclear emulsion has been built (Fig. 19). In a trial operation, they have achieved continuous coating and drying of 0.5 m/min (0.15 m²/min) with a coating width of 0.3 m. As for the thickness of the coating film, a coating thickness of 40–80 μm and a coating accuracy of 1.2–1.8 μm (for coating thicknesses of 40 μm and 80 μm, respectively) have been achieved. When operated at this speed, 1,000 m² (both sides) per year can be achieved in approximately 40 days of operation at 6 h/day.

On the other hand, in 2015, it was reported that production of nuclear emulsions with characteristics required for relativistic particle registration was organized at the Slavich OJSC in Russia.[37] At present, a supply of nuclear emulsions is being organized to meet various needs, including muon radiography.

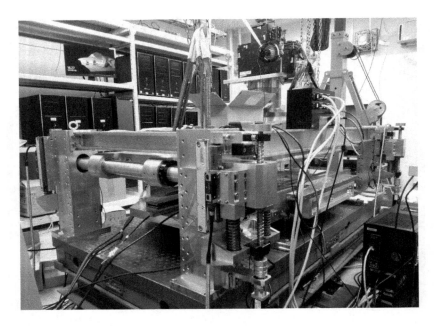

Figure 20. Next generation scanning system HTS2 under construction at Nagoya University.

3.4. *Scanning speed*

As already described, the current fastest scanning system, HTS,[30] achieves a reading speed of $0.45\,\mathrm{m^2/h}$. In the HTS, the field of view of the objective lens is enlarged, and 72 sensors are used to capture a wide field of view at once, resulting in higher speed. In the next stage of the system, HTS2 (Fig. 20), they aim to achieve a scanning speed of $2.5\,\mathrm{m^2/h}$ by expanding the field of view of the microscope and by performing continuous reading without stopping the stage. In addition, the 12-fold wider moving range of the x-y stage is expected to reduce the time required for nuclear emulsion replacement. Table 1 shows performance comparison of HTS2 with HTS.

4. Summary

Emulsion films have been considerably important in muon radiography since the dawn of the 2000s. As X-ray radiography technology has

Table 1. Performance comparison of HTS2 with HTS.

Machine	Start of operation	x-y stage motion range	Number of image sensors	Read rate	Scanning speed
HTS	2014	$12.5 \times 10 \, \text{cm}^2$	72	5 Hz	$0.45 \, \text{m}^2/\text{h}$
HTS2	2021 (schedule)	$40 \times 37.5 \, \text{cm}^2$	72	continuous	$2.5 \, \text{m}^2/\text{h}$

moved from film to digital detectors, the role of electrical detectors will increase in the future of muon radiography, particularly in monitoring applications. However, emulsion films will continue to be significant for sites where there is no power supply, sites where large detectors are inaccessible, and for one-time measurements and simultaneous deployment in multiple locations.

In this chapter, we described some of the developments of nuclear emulsion detectors in muon radiography. The nuclear emulsion detector, which has solved the problem of long-term stability and enabled mass production and high-speed readout, will further expand the scope of muon radiography technology in the future.

References

1. A. Ariga, T. Ariga, G. De Lellis, A. Ereditato, and K. Niwa, Nuclear emulsions. In *Particle Physics Reference Library*, pp. 383–438. Springer (2020).
2. H. Tanaka, T. Nakano, S. Takahashi, J. Yoshida, H. Ohshima, T. Maekawa, H. Watanabe, and K. Niwa, Imaging the conduit size of the dome with cosmic-ray muons: The structure beneath Showa-Shinzan Lava Dome, Japan, *Geophysical Research Letters*, **34**(22), (2007).
3. H.K.M. Tanaka, T. Nakano, S. Takahashi, J. Yoshida, M. Takeo, J. Oikawa, T. Ohminato, Y. Aoki, E. Koyama, H. Tsuji, *et al.*, High resolution imaging in the inhomogeneous crust with cosmic-ray muon radiography: The density structure below the volcanic crater floor of mt. asama, japan, *Earth and Planetary Science Letters*, **263**(1), 104–113 (2007).
4. A.H. Becquerel, On the invisible rays emitted by phosphorescent bodies, *Comptes Rendus*, **122**, 501–503 (1896).

5. S. Kinoshita, The photographic action of the α-particles emitted from radio-active substances, *Proceedings of the Royal Society of London. Series A, Containing Papers of a Mathematical and Physical Character*, **83**(564), 432–453 (1910).
6. Reinganum, Phys, *Zeit.* **xii**, 1076–1077 (1911).
7. C.M.G. Lattes, H. Muirhead, G.P.S. Occhialini, and C.F. Powell, Processes involving charged mesons, *Nature*, **159**(4047), 694 (1947).
8. K. Niu, E. Mikumo, and Y. Maeda, A possible decay in flight of a new type particle, *Progress of Theoretical Physics*, **46**(5), 1644–1646 (1971).
9. K. Hoshino, N. Ushida, K. Niwa, Y. Maeda, S. Kuramata, K. Niu, E. Mikumo, and S. Tasaka, X-particle production in 205-gev/c proton interactions, *Progress of Theoretical Physics*, **53**(FERMILAB-PUB-75-150-E), 1859 (1975).
10. S. Aoki, R. Arnold, G. Baroni, M. Barth, J. Bartley, G. Bertrand-Coremans, V. Bisi, A. Breslin, G. Carboni, E. Chesi, *et al.*, The double associated production of charmed particles by the interaction of 350 gev/cπ-mesons with emulsion nuclei, *Physics Letters B*, **187**(1–2), 185–190, (1987).
11. E. Eskut, A. Kayis-Topaksu, G. Önengüt, M. van Beuzekom, R. van Dantzig, M. de Jong, J. Konijn, O. Melzer, R.G.C. Oldeman, E. Pesen, *et al.*, Final results on $\nu\mu$ $\nu\tau$ oscillation from the chorus experiment, *Nuclear physics B*, **793**(1–2), 326–343, (2008).
12. K. Kodama, N. Ushida, C. Andreopoulos, N. Saoulidou, G. Tzanakos, P. Yager, B. Baller, D. Boehnlein, W. Freeman, B. Lundberg, *et al.*, Observation of tau neutrino interactions, *Physics Letters B*, **504**(3), 218–224, (2001).
13. K. Kodama, N. Ushida, C. Andreopoulos, N. Saoulidou, G. Tzanakos, P. Yager, B. Baller, D. Boehnlein, W. Freeman, B. Lundberg, *et al.*, Final tau-neutrino results from the donut experiment, *Physical Review D*, **78**(5), 052002 (2008).
14. R. Acquafredda, N. Agafonova, M. Ambrosio, A. Anokhina, S. Aoki, A. Ariga, L. Arrabito, D. Autiero, A. Badertscher, E. Baussan, *et al.*, First events from the cngs neutrino beam detected in the opera experiment, *New Journal of Physics*, **8**(12), 303 (2006).
15. N. Agafonova, A. Aleksandrov, A. Anokhina, S. Aoki, A. Ariga, T. Ariga, D. Bender, A. Bertolin, I. Bodnarchuk, C. Bozza, *et al.*, Discovery of τ neutrino appearance in the cngs neutrino beam with the opera experiment, *Physical review letters*, **115**(12), 121802 (2015).
16. Myssowsky and Tschishow, *Z. Phys.*, **44**, 408, (1927).
17. M. Blau and H. Wambacher, Photographic desensitisers and oxygen, *Nature*, **134**(3388), 538, (1934).

18. R. Berriman, Recording of charged particles of minimum ionizing power in photographic emulsions, *Nature*, **162**(4130), 992–993 (1948).
19. R. Brown, U. Camerini, P.H. Fowler, H. Muirhead, C.F. Powell, and D.M. Ritson, Observations with electron-sensitive plates exposed to cosmic radiation, *Nature*, **163**(4132), 47, (1949).
20. T. Nakamura, A. Ariga, T. Ban, T. Fukuda, T. Fukuda, T. Fujioka, T. Furukawa, K. Hamada, H. Hayashi, S. Hiramatsu, *et al.*, The opera film: New nuclear emulsion for large-scale, high-precision experiments, *Nuclear Instruments and Methods in Physics Research Section A: Accelerators, Spectrometers, Detectors and Associated Equipment*, **556**(1), 80–86, (2006).
21. K. Kuwabara and S. Nishiyama, Development of new nuclear emulsion film for detection of neutrinos by opera experiment, *Journal of The Society of Photographic Science and Technology of Japan*, **67**(6), 521–526 (in Japanese), (2004).
22. A. Nishio, K. Morishima, K. Kuwabara, T. Yoshida, T. Funakubo, N. Kitagawa, M. Kuno, Y. Manabe, and M. Nakamura, Nuclear emulsion with excellent long-term stability developed for cosmic-ray imaging, *Nuclear Instruments and Methods in Physics Research A* (submitted).
23. C. Amsler, A. Ariga, T. Ariga, S. Braccini, C. Canali, A. Ereditato, J. Kawada, M. Kimura, I. Kreslo, C. Pistillo, *et al.*, A new application of emulsions to measure the gravitational force on antihydrogen, *Journal of instrumentation*, **8**(02), P02015, (2013).
24. K. Niwa, K. Hoshino, and K. Niu, Auto scanning and measuring system for the emulsion chamber, *the proceedings of the International Cosmic ray Symposium of High Energy Phenomena, Tokyo*, **149** (1974).
25. S. Aoki, K. Hoshino, M. Nakamura, K. Niu, K. Niwa, and N. Torii, Fully automated emulsion analysis system, *Nuclear Instruments and Methods in Physics Research Section B: Beam Interactions with Materials and Atoms*, **51**(4), 466–472 (1990).
26. T. Nakano, Ph.d. thesis, *Nagoya University* (1997).
27. K. Morishima and T. Nakano, Development of a new automatic nuclear emulsion scanning system, S-UTS, with continuous 3d tomographic image read-out, *Journal of Instrumentation*, **5**(04), P04011, (2010).
28. N. Armenise, M. De Serio, M. Ieva, M. Muciaccia, A. Pastore, S. Simone, J. Damet, I. Kreslo, N. Savvinov, T. Waelchli, *et al.*, High-speed particle tracking in nuclear emulsion by last-generation automatic microscopes, *Nuclear Instruments and Methods in Physics Research Section A: Accelerators, Spectrometers, Detectors and Associated Equipment*, **551**(2–3), 261–270, (2005).

29. L. Arrabito, E. Barbuto, C. Bozza, S. Buontempo, L. Consiglio, D. Coppola, M. Cozzi, J. Damet, N. D'Ambrosio, G. De Lellis, *et al.*, Hardware performance of a scanning system for high speed analysis of nuclear emulsions, *Nuclear Instruments and Methods in Physics Research Section A: Accelerators, Spectrometers, Detectors and Associated Equipment*, **568**(2), 578–587, (2006).

30. M. Yoshimoto, T. Nakano, R. Komatani, and H. Kawahara, Hyper-track selector nuclear emulsion readout system aimed at scanning an area of one thousand square meters, *Progress of Theoretical and Experimental Physics*, **2017**(10), (2017).

31. A. Alexandrov, A. Buonaura, L. Consiglio, N. D'Ambrosio, G. De Lellis, A. Di Crescenzo, G. Galati, V. Gentile, A. Lauria, M.C. Montesi, *et al.*, The continuous motion technique for a new generation of scanning systems, *Scientific Reports*, **7**(1), 1–10, (2017).

32. A. Alexandrov, G. De Lellis, and V. Tioukov, A novel optical scanning technique with an inclined focusing plane, *Scientific Reports*, **9**(1), 1–10, (2019).

33. K. Morishima, M. Kuno, A. Nishio, N. Kitagawa, Y. Manabe, M. Moto, F. Takasaki, H. Fujii, K. Satoh, H. Kodama, *et al.*, Discovery of a big void in Khufu's Pyramid by observation of cosmic-ray muons, *Nature*, **552**(7685), 386, (2017).

34. H. Tanaka, T. Nakano, S. Takahashi, J. Yoshida, and K. Niwa, Development of an emulsion imaging system for cosmic-ray muon radiography to explore the internal structure of a volcano, mt. asama, *Nuclear Instruments and Methods in Physics Research Section A: Accelerators, Spectrometers, Detectors and Associated Equipment*, **575**(3), 489–497, (2007).

35. N. Agafonova, A. Aleksandrov, O. Altinok, A. Anokhina, S. Aoki, A. Ariga, T. Ariga, D. Autiero, A. Badertscher, A. Bagulya, *et al.*, Momentum measurement by the multiple coulomb scattering method in the opera lead-emulsion target, *New Journal of Physics*, **14**(1), 013026, (2012).

36. R. Nishiyama, A. Taketa, S. Miyamoto, and K. Kasahara, Monte carlo simulation for background study of geophysical inspection with cosmic-ray muons, *Geophysical Journal International*, **206**(2), 1039–1050, (2016).

37. A. Aleksandrov, A. Bagulya, M. Chernyavsky, N. Konovalova, A. Managadze, O. Orurk, N. Polukhina, T. Roganova, T. Shchedrina, N. Starkov, *et al.*, Muon radiography in russia with emulsion technique. first experiments future perspectives. In *AIP Conference Proceedings*, vol. 1702, pp. 110002–1, AIP Publishing LLC (2015).

© 2023 World Scientific Publishing Company

https://doi.org/10.1142/9789811264917_0005

Chapter 5

Real-time Detectors for Muography

F. Ambrosino* and G. Saracino[†]

Università degli Studi di Napoli Federico II
Dipartimento di Fisica "Ettore Pancini"
and
Istituto Nazionale di Fisica Nucleare, Sezione di Napoli
Via Cintia 80126 Napoli — IT
*fabio.ambrosino@unina.it
[†]giulio.saracino@unina.it

Different types of detectors can be used for muon radiography. Common to all detectors is the ability to *track* the direction of the incoming muon. Emulsion detectors have very high resolution and no need for electric power, but cannot give real-time information and cannot be used in all environmental conditions. Detectors with electronic readout have generally lower resolution but are more flexible, can be operated in real-time, and can also measure the time of the muon hitting the detector with resolutions as low as few hundred ps in specific setups. Special geometries and assemblies have been used for specific applications, where the size, weight, or shape of the detector are heavily constrained.

1. Real-time Detectors

The detection and tracking of muons can be done adapting to this purpose various types of tracking detectors originally conceived for particle physics experiments. Historically, the first-ever high resolution muon radiography, performed by L. Alvarez inside the Chefren Pyramid,[1] used a gas-based (streamer tube) electronic detector. Unlike emulsion detectors, gas detectors and scintillator-based trackers can be read out in real-time using electronic readout

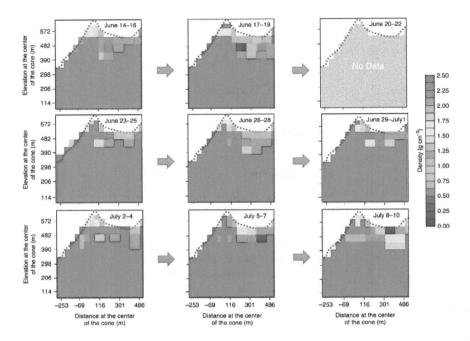

Figure 1. Sequence of muographic images of Satsuma-Iwuojima volcano in Japan during about 1 month of data taking. From Ref. 3.

and data storage systems, which may be highly automatized and remotely controlled. Modular units of electronic detectors may be assembled to form large area detectors in order to improve sensitivity to low muon fluxes (i.e., large rock thicknesses) with no increase in data processing time. Unique to these detectors is the possibility to sample the muon radiography, either real-time or offline, in time windows of any desired duration (limited though by the muon flux) and to observe time variations in the features of the radiography itself (see for instance Fig. 1, from Refs. 2,3).

Detectors of this type are flexible, configurable, re-usable, and can be operated in the widest range of temperatures in applications on field; on the other hand, they require some source of electric power for their operation. Moreover, in order to reach the required angular resolution for tracks, these detectors are inherently less compact than emulsion-based detectors because of their worse single-point spatial resolution. Optimization of power consumption, development of robust detectors and readout electronics, and optimization of single-point resolution are thus key issues in developing these kind of devices for muon radiography.

There are some principles common to all electronic detectors. The passage of a muon across a material is accompanied by a continuous energy release, which, depending on the material chosen, can result in a *physical signal* in the form of excitation of electrons inside the molecule/lattice (followed by de-excitation via emission of scintillating light), or in the form of gas ionization. These energy releases along the muon path are sampled in real-time by an appropriate array of sensors (photodetectors, anode wires/strips, etc.), which convert the *physical signal* into an *analog electronic signal*, i.e., a current pulse usually having a fast rising time and a short duration. Finally, these pulses are converted by the read-out electronics into *digital signals*, i.e., (series of) numbers, and stored in digital form on mass-memory systems for subsequent offline elaboration. Typical offline elaboration includes application of inter-calibration correction among different electronic channels, noise suppression, track recognition, and reconstruction of trajectories and (for some detectors) time of passage of the muon through the detection system. A conceptual scheme of a detection element is sketched in Fig. 2.

Despite these common features, different types of electronic detectors will usually have different performances and may be optimized for specific applications, as we will see in the following.

Application of muon radiography to larger and larger targets is a big challenge; however, the study of large volcanic edifices requiring exploration of rock thicknesses at the km scale are currently planned or ongoing. The use of electronic detectors is necessary in these cases for the detection of lower and lower incoming muon fluxes, which require increase in acceptance and reduction of backgrounds. As already discussed in Chapter 3, backgrounds can originate from low energy muons, which have scattered through the superficial part of the target and have lost their original direction, as well as from gamma rays and extended showers which may cause simultaneous hits in the detector planes, faking a muon track. Reduction of these

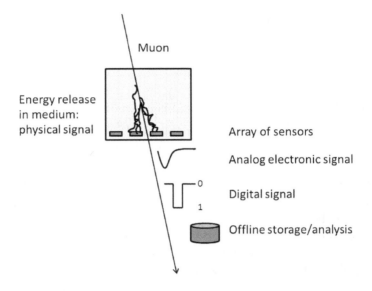

Figure 2. Conceptual scheme of an element of an electronic detector for muon radiography.

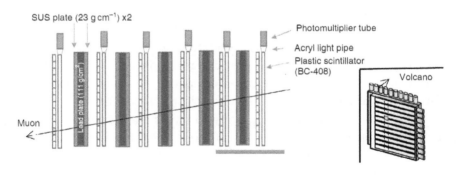

Figure 3. Schematic view of a multi-layer muon detection system, from Ref. 3.

backgrounds is achieved by increasing redundancy in the number of detection planes (it is now common to use at least 3 or 4 planes for this purpose, and even more[3] have been used in some applications) and by using shielding (a high Z material layer such as lead) in between the detection planes to stop low-energy particles, see Fig. 3

for example. A new concept in this sense is the one using large area Cherenkov telescopes, as we will discuss briefly in the following.

On the opposite side, in archaeological and civil use of muon radiography it is often important for logistic problems to keep detector size at a minimum, while a higher level of background is usually tolerable given the lower amount of material (typically some tens of metres) to be explored by the muography. The development of compact detectors is then highly advisable, and while emulsion detectors are clearly the most compact detector available, electronic detectors have been developed and used also for these purposes, with a good compromise between the conflicting needs for angular resolution and compact size.

1.1. *Scintillator-based detectors*

1.1.1. *Scintillation process*

When traversing a dense material, a muon (as any charged particle) interacts with the medium and loses a small fraction of its energy in the excitation of the electrons of the molecules (or the lattice) which compose the material. This energy is then released in the form of emitted "scintillation" light after a de-excitation time specific of the material itself. Details on the various different mechanisms generating scintillation and fluorescence light as well as a review of different types of scintillators can be found elsewhere,[4-6] here we will limit ourselves to describing some very basic ideas relevant to muon radiography. Scintillation light can be observed both in *inorganic crystals* and in *organic (plastic)* materials. Given their light weight, their low cost, their robustness with respect to operating conditions and their fast response, plastic scintillators are the typical choice in muon radiography. If special care is taken in the chemical composition of the material, also using so-called "dopants", (which effectively act as wavelength-shifters, absorbing the original light and re-emitting it at a larger wavelength) the bulk of a plastic scintillator may be made transparent to part of the emitted light, whose wavelength is shifted from the original UV to a typical value around 400 nm (violet). A photodetector is then eventually able to catch this light and transform it into an electronic pulse. The process of "seeing" a muon with a scintillator detector involves thus a multi-step approach: energy deposit in the medium, light emission from the scintillator, and collection of this light onto the photodetector active surface, finally conversion of this light into electronic signal by the photodetector itself. Each of these steps involves some losses, and maximizing efficiency at each step is crucial if one wants to keep the dimensions and weight of the detector to a minimum. In order to set the scale, we observe that a minimum ionizing particle crossing a 1 cm deep plastic scintillator will produce $O(10^4)$ primary photons, of which, unless the detector is very finely segmented, only a fraction (which can be of the order of 10–20%, depending on geometry and treating of surfaces) will eventually reach the sensitive area of the photodetector being converted (again with efficiencies of order 10–30%) to form the electronic signal. Thus, in order to have a reasonable detection efficiency, each plane/view of a scintillator-based detector cannot be made too thin: a two centimeters deep view will weigh $O(20\,\text{kg/m}^2)$ and the weight of a scintillator-based detection plane equipped with two views, including support mechanics and cables, will easily reach $O(50–100)\,\text{kg/m}^2$.

1.1.2. *Photodetection*

Photodetection is an important part of the process of measuring muons with scintillators. Since typically the amount of light to be detected is very small (sometimes as low as ten photons) special devices must be used to this aim. A typical system to detect scintillation light, widely used in particle physics, is via the photoelectric effect and subsequent amplification of the signal via a cascade of high voltage electrodes put inside a vacuum tube (dynodes). A large variety of *photomultiplier tubes* (PMTs) has been developed for this purpose, and may be produced with different values for the area of the sensitive surface, with optimized sensitivity in different regions of the light spectrum and with various levels of intrinsic amplification. This is a very well consolidated technique used for almost a century in particle physics experiments. More recently, a different class of photodetectors has become available. These devices, called usually Silicon Photomultipliers (for a recent review see e.g.,[7] and references therein) or SiPMs but known also with different names depending on the manifacturer, use a completely different mechanism to generate the electronic signal, i.e., the creation of an electron–hole couple inside the depletion region of a reverse-biased semiconductor and the subsequent cascade amplification of the signal in the same material. Unlike photomultiplier tubes, SiPMs are "digital" devices in the sense that the sensitive surface is divided into pixels which have only a yes/no response to the passage of a photon. However, as long as enough pixels are present per unit surface (typical values are of order of few hundred pixels/mm^2 of active surface) the number of fired pixels is usually proportional to the number of photons which originally hit the SiPM surface as the latter is far from saturating the response of the device. PMTs and SiPMs have different pros and cons for a muography application. PMTs require high voltage (typically from few hundred to few thousand Volts) for the multiplication stages, while SiPMs only require voltages at level of some tens of Volts. Power consumption is lower for SiPMs with respect to PMTs. From the mechanical point of view, SiPMs are more robust and compact objects (though they usually have a much smaller sensitive area). On the other hand, SiPMs are quite sensitive to ambient temperature variations, which may change their operational parameters and their level of noise. Moreover, given the limitations on the area of the sensitive surface, SiPM-based systems match well with a finer segmentation for the detector (which however requires a large number of devices and electronic channels). A finer segmentation is associated to better resolution in position and tracking, but of course the larger number of electronic channels introduces a complication for the system which may not be necessary for low resolution applications. It must be said, however, that some of the current limitations related to SiPMs are going to become less and less relevant with time since the development of SiPMs is an on-going process, and devices with lower and lower noise and increasingly larger active surface become available on the market. New generation devices, commercially available since the late 2010s, have in fact decreased by more than an order of magnitude their intrinsic noise with respect to first generation commercial devices. The so-called dark rate, i.e., the rate of "1 p.e." signals (counted setting a threshold at half the single photo-electron signal peak amplitude) produced at room temperature by thermal fluctuations in absence of a photon hitting the device, is now typically at level $\approx 50 \, \text{kHz/mm}^2$. Tripling the threshold this rate scales down by more than one order of magnitude thanks to the improvements in multi-photon noise (i.e., reduction of "optical cross talk", see e.g.,[7] for details). If enough care in light yield of the scintillator and collection efficiency is taken, the signal corresponding to a muon will be significantly larger than the "1 p.e." amplitude, allowing to set higher thresholds and thus strongly suppress this kind of noise.

1.1.3. *Examples*

An early example of scintillator-based detector for muon radiography is the one used in 2008 to study the Satsuma-Iwojima Volcano, in Japan.[2] The detector was very simple in its structure: 24 bars of scintillators 1500 mm long and 78 mm wide, were arranged in two orthogonal views to form a matrix, with a $\approx 8 \times 8 \, \text{cm}^2$ "pixel" defined by the region of superposition of two orthogonal bars. Each bar was in turn readout using a PMT (Hamamatsu H7724). The electronic coincidence of signals within a 200 ns time window were accumulated in bins of their incoming direction angles into a 2D histogram acquired using a customized Field Programmable Gate Array, which was also capable of sending data in HTML form to the remote mass-memory system. Two planes were used to track the direction of the muon. The total weight of the 48 scintillator modules was aboout 163 kg. The system allowed to reach an angular resolution of 14 mrad by setting the two planes at 2 m distance, and since the detector was placed at 1.5 km from the volcano, the spatial resolution for density anomalies and structures of the volcano was 17 m. This kind of detector was the first to show clearly the potential of real-time muon radiography in the study of volcanic edifices and other

Figure 4. Left: one MU-RAY module before and after encasement. Right: 12 MU-RAY modules assembled to form a three-planes, 1 m² detector. From Ref. 9.

geophysical targets.[8] However, given the relatively large pixel size, and hence the poor single-point spatial resolution, its acceptance was limited by the need to put the two detection planes at 2 m distance in order to have the desired angular resolution.

A step ahead to a new generation of scintillator-based detectors for muography was reached by the MU-RAY R&D project[9,10] and the subsequent MURAVES large-area telescope[11,12] built to study the Vesuvius in Italy. Figure 4 shows the detector structure.

The detector was designed to reach a single-point resolution an order of magnitude better than the one obtained by first-generation scintillator telescopes. This was achieved using special 1 m long scintillator bars, with sections in the shape of an isosceles triangle (base 33 mm, height 17 mm) arranged with their vertices alternatively upside-down to form a plane. Each bar was read by a wavelength shifting optical fiber which was in turn coupled to a SiPM. Particular care was taken in all the optical couplings in order to minimize light losses; the wavelength-shifting fiber diameter was optimized to match the SiPM sensitive surface. The geometry of such a system is such that an incoming muon, traversing almost at right angle the detection plane will cross two bars and its path length in each bar will decrease linearly with increasing distance from the bar center. Since the energy release of the incoming muon, and hence the amount of light detected by each SiPM, is proportional to the path length of the muon inside the bar, the amount of light collected by each bar will decrease linearly from a maximum when the muon hits the center of the bar, down to almost zero when it hits one of the two extremes of the triangle base. Thus, the amount of light collected by each bar can be used as a weight for a weighed average determination of the impact point. The situation is depicted in Fig. 5.

In turn this situation allows to reach a single-point spatial resolution better than the one expected from the pitch of the bar centers (i.e., $16.5\,\mathrm{mm}/\sqrt{12} = 4.7\,\mathrm{mm}$). In fact, spatial resolution down to $\sim$3 mm have been reported using this geometry. Since the granularity of the detector is high, this system has to be equipped with a relatively large number of electronic channels: for a $1\,\mathrm{m}^2$ and 2 views the number of photodetectors and readout channels is 128 (to be compared with 24 in the detector described above[8]). Since SiPMs require a stabilization in temperature to be operated optimally, a thermoregulation system based on Peltier cells has been developed for MURAVES, which allows to adjust temperature within 5°C of the ambient temperature with a consumption of less than 5 W/layer. An important feature of the MURAVES detector is that since the whole scintillation process is fast (typical light-emission

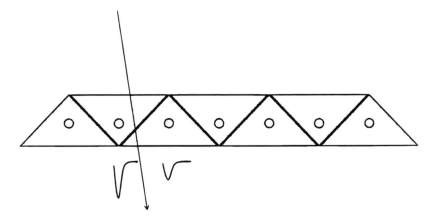

Figure 5. Scintillator bars of triangular section used in MU-RAY/MURAVES: the energy release, measured by the electronic signal, is proportional to the path of the muon inside the bar.

times of few ns), the time of a muon hit can in principle be measured with a resolution of O(ns). One can then distinguish the direction of motion of the incoming muons: in fact, the muon, travelling at a speed very close to c, will hit the first plane along its path about 3–6 ns before hitting the last plane, them being separated by a distance of order 1–2 m. Unambiguous determination of the flight direction needs a time resolution of O(ns), which can be reached using fast scintillators and appropriate readout electronics.

Scintillator detectors are easily arranged in modular substructures, which helps transporting and installing them on-field. For instance, the MURAVES detector is assembled starting from modules made up of 32 bars, which cover approximately a $1 \times 0.5 \, \text{m}^2$ surface for one view. Since four planes, $1 \, \text{m}^2$ each, are used for each telescope, a total of 16 modules are needed to assemble a telescope; while the overall system is quite heavy, each module has still a weight and dimensions acceptable for being handled and transported by two people if necessary.

Ruggedization and insensitivity to water and other hard environmental conditions is also desirable on field, and scintillation detectors are well-suited to this purpose. For instance, a muon radiography of La Soufrière of Guadeloupe volcano has been obtained in the context of the DIAPHANE project[13] using a scintillator detector[14] made of 32 bars/plane with a pixel size of $5 \times 5 \, \text{cm}^2$, which had been carefully designed to be immune to the harsh environmental conditions in which it had to operate (see Fig. 6).

The system was also optimized for low power consumption (40 W total), and it was able to work in continuous operation even in cloudy and rainy weather conditions using a system of solar panels providing a peak power of 720 W.

1.2. *Gas detectors*

Gas detectors are the preferred choice for tracking charged particles far from the interaction region in most particle physics experiments. They are light-weight, and can reach very good single-point resolution with respect to typical scintillator trackers: in modern detectors,

Figure 6. The muon radiography detector at La Soufrière (from Ref. 13).

single-point resolution better than 100 μm can be obtained. Some of these devices are commonly used to track muons in the outer shells of the large-sized particle detectors at accelerators (see e.g.,[15, 16]), and optimized to cover large surfaces at reasonable costs. These features allow to consider the gas detectors as a good choice in perspective for muon radiography if one wants to maximize acceptance (exploiting the large size and the possibility to keep a good angular resolution also with tracking planes placed at a relatively close distance) or to reduce size (exploiting the good resolution to minimize the detector size). The price to pay for this performance is the need for a gas system, which either periodically purifies (closed mode operation) or replaces (open mode operation) the gas used by the detector and a mechanical complexity (wires, meshes etc.) which is usually a concern if the system has to operate in harsh environmental conditions. Another important issue is that gas properties are sensitive to ambient temperature and pressure, and thus the detector operating parameters may need to be properly adjusted to follow environmental variations with time.

The principle of operation of a gas detector is the following: the high momentum charged particle (in our case, the muon), while traversing a gas volume, will continuously ionize the gas itself along its path, producing a certain number of electron-ion couples. The gas is typically immersed in a strong electric field generated by apposite electrodes, so that electrons drift toward the anode and in this process generate further ionizations (*multiplication* of the signal) leading to the development of an electron avalanche. Depending on geometry and choice of the multiplication mechanisms, the original charge (of order $5Ze/cm$ in a gas with atomic number Z) is usually multiplied by several (usually five or six) order of magnitudes, so that when collected at the anode it is significantly higher than the typical electronic noise (usually of order $1,000\ e$) and can be registered and converted into digital information. One takes care to operate in the so-called *proportional* regime, where the charge of the avalanche collected at the anode is in fixed proportion to the primary charge if the amount of energy lost by the muon has to be measured. More often, one only needs a fast yes–no information on the passage of the muon, and in this case the detector can operate in the so-called streamer regime, in which this fixed proportion is lost. Reader interested in detailed operation of gas detectors can refer to the vast literature on the subject, starting from classical textbooks[4, 5] and references therein, or referring to the most up-to-date reviews.[6] Spatial information on the passage of the muon may be obtained by aligning several anode sensors (tubes, wires, strips...) along one direction (say x) at close pitch: the identification of the sensor fired will allow to measure the position of the hitting particle along the direction (say y) orthogonal to the tube/wire/strip. A combination of detectors with tubes/wires/strips along orthogonal directions will form the usual two-views plane to measure the position of the hitting muon. More complex geometries may be used (use of three or more non-orthogonal wires for redundancy, use of pads/pixels instead of strips, etc.) and a variety of different gas mixtures (usually composed of a noble gas and a small amount of an organic molecule used as an absorber to prevent UV light generated in the multiplication phase to start avalanches far from the original electrode) can be adjusted to obtain the desired performances of the detector. When working in so-called *closed mode*, i.e., with a fixed amount of gas trapped inside the detector, impurities and pollutants leaking from external air, and/or generated by electrochemical processes inside the detector tend to accumulate inside the sensitive volume, degrading the detection performances: for this reason either a periodic purification of the gas, or a complete flush and replacement of it is needed for long time operation. This adds of course some logistic difficulties to the operation of gas detectors for muon radiography on the field.

1.2.1. *Examples*

The first detector used for a muon radiography by L. Alvarez was a gas detector. The tracker used so-called streamer-tubes to measure the passage of the muon and integrated the incoming direction in bins of a 2D histogram. The detector was able to clearly resolve the edges of the Chefren pyramid showing the potential of a high-resolution tracker in identifying relatively small-scale structures with muon radiography.

More recently, the TOMUVOL collaboration[17] operated a detector based on Glass Resistive Plate Chambers[18] (GRPCs) and used it to perform muon radiography of the Puy de Dome volcano in France.[19] Resistive Plate Chambers (RPCs) are gaseous detectors where the intense electric field needed for avalanche multiplication of the primary ionization is achieved by means of planar electrodes placed at very close distance (1.2 mm in this case) and high voltage (in this case about 7 kV) surrounding the gas volume. A thin highly resistive plate (bakelite, or graphite covered glass for GRPCs) separates the gas from the electrodes, usually strips or pads (1 cm^2 pads were used by TOMUVOL) and ensures that the induced electric signal is contained in the vicinity of the passage of the particle. A schematic representation of this detector is shown in Fig. 7.

These detectors may be produced and assembled to cover large surface areas and have a low cost/unit surface. TOMUVOL's GRPCs reached a single hit resolution of 0.4 cm, though using a very large number of readout electronic channels (9,472/1 m^2 plane, for a total

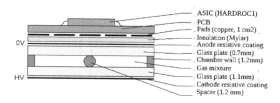

Figure 7. Schematic representation of a planar Glass Resistive Plate Chamber (from Ref. 18).

of 37,888 channels for a 4-planes detector). The detector was operated in a laboratory in fluxed open mode, with full replacement of the gas mixture achieved about every two hours. It is important to point out that, unlike some other gas detectors, GRPCs are intrinsically very fast devices, which produce signals on a O(ns) time scale; so that if read by an appropriate electronics they can also provide the time of flight of the muon, like fast scintillators.

1.2.2. *Cherenkov detectors*

Very high energy charged particles crossing a medium with an index of refraction n may occasionally have a speed v larger than the light speed in the medium (c/n). In this situation, a phenomenon analogous to the sound shock wave generated by jets moving in air faster than sound occurs in the medium, and a so-called Cherenkov light emission occurs. Cherenkov light is concentrated on the surface of a cone whose axis is along the direction of motion of the particle, and whose aperture 2θ is given by $\cos\theta = \frac{c}{nv}$ (see Fig. 8).

Cherenkov light is widely used in particle physics to detect high energy charged particles and measure their velocity, by imaging the ring generated by the Cherenkov cone on a given sensitive surface. The ASTRI collaboration has proposed to use this approach to perform a muon radiography of a volcano, by using a gamma ray telescope to measure the Cherenkov ring generated by muons surviving after having crossed the volcanic edifice.[20] The advantage of this technique is that the suppression of low-energy particle is

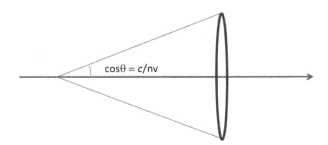

Figure 8. Cherenkov light emission is essentially confined on the surface of a cone along the path of the charged track.

achieved automatically, since the minimum energy for a muon to produce Cherenkov radiation in air is about 5 GeV. However, data-taking must be limited to night periods, in order for the telescope not to be blinded by ambient light; costs/unit surface and portability of these kind of detectors, which are not intrinsically modular, also appear as a potential issue for muon radiography applications on field.

1.3. *Portable devices*

Application of muography to volcanoes and other geological targets requires to maximize acceptance given the low fluxes, and implies detectors with a large sensitive area. This may not be optimal for detectors aiming at different applications, like archaeology, civil engineering, etc. Here, typical rock overburdens/material thicknesses may be of "only" few tens of metres, so that a compact, portable device may suffice to detect a density anomaly in a reasonable amount of time, and may thus be preferable for obvious logistic reasons. Ideally, in order to keep a good angular resolution a compact device should have a very good single-point resolution; however, in practice since anomalies are usually observed relatively close to the detector, and hence under a wide solid angle, this parameter turns out to be less crucial than expected. While most detectors for muon radiography of volcanoes are designed as modular assemblies, so that a smaller assembly can be used as a portable device (for instance MU-RAY/MURAVES modules have been used also for these

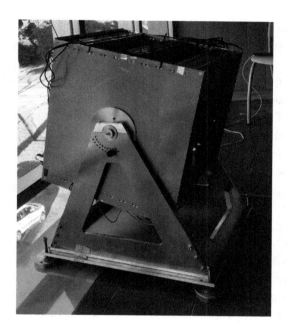

Figure 9. The MIMA detector with its alt-azimuth mount (from Ref. 23).

purposes[21, 22]) some muon radiography detectors have been conceived specifically for being portable and small-sized.

The MIMA project[23] is a scaled-down version of the MU-RAY/ MURAVES detector described above. It is compact ($50 \times 50 \times 50 \, \text{cm}^3$, see Fig. 9), relatively lightweight (about 60 kg including electronics), and has low power consumption (about 30 W). Following developments in scintillator and SiPM technologies, for this detector no wavelength shifting fiber was used, thus simplifying the construction technique. Single-point resolution of 3.3 mm and angular resolution of 14 mrad has been achieved for this portable detector, which can be used also in positions where the MU-RAY/MURAVES modules (whose dimensions are of $O(1 \, \text{m}^3)$) can't fit.

Plastic scintillating fibers are plastic scintillators shaped in the form of a fiber, having appropriate refractive index and cladding so as to act also as light-guides for the emitted photons. In Ref. 24 the authors used 200 mm long fibers (with a square section of $2 \times 2 \, \text{mm}^2$

each) bundled in groups of four and coupled to SiPMs to build a compact and relatively low-power consumption (155 W) detector with the usual xy arrangement and two detection planes. This device was specifically designed to be portable, in order to be used for infrastructure degradation monitoring. A satisfactory angular resolution of 8 msr was achieved, allowing the muographic imaging of concrete thicknesses of $\simeq$1,000 cm deep, with typical uncertainties below 100 cm.

Along the lines of development of compact detectors, an extremely small and lightweight gas-based system[25] has been developed in 2012. This system was enclosed in a $51 \times 46 \times 32$ cm^3 envelope with a weight of only 13 kg, and was based on a design similar to the one of conventional Multi-wire Proportional Chambers, called Close Cathode Chamber (CCC). Multi-wire Proportional Chambers (MWPC) are gas detectors which use arrays of sensitive anode wires in between two cathode planes. While commonly used in "digital" yes–no response, since they work in proportional regime, the amplitude of the response of adjacent sense wires may be used as information for a "center of charge" technique (in a way analogous somehow to the mechanism described above for a triangular shaped scintillator) and can achieve a position resolution along the coordinate orthogonal to the wires which is smaller than the one given by the wire pitch. The CCC is essentially an asymmetric MWPC where one of the two cathode planes is placed very close (1.5–2 mm distance) to the sense wires, and is itself equipped with sensitive strips (4 mm wide) along the direction orthogonal to the wires. In this way, each planar assembly (1 cm deep) gives information on the 2D position of the track hitting the detector, with a resolution of about 1.5 mm. Considering the geometry of the detector planes (two couples of 32×32 cm^2 planes at about 16 cm distance) this results in a satisfactory angular resolution of about 10 mrad. The whole system used in[25] was closed in a protective Al-Plexiglas box, which isolated the detector from the environment, and allowed to operate the detector in humid conditions (caves) and/or close to the dew point. A gas system was needed to continuously flux the gas mixture (Ar-CO$_2$ 80%–20%):

for underground measurements in the Ajàndék Cave, a 100 m long rubber tube was used to supply gas (about 3 l/h) from an external bottle, which was able to ensure continuous operation for about 20 days. The exhaust gas flow was recycled inside the protective box to ensure a low humidity local environment, where RH of 30–50% was measured. The device is shown in Fig. 10.

Another example of a gas-based detector, with good transportability and high spatial resolution, was developed at CEA-Saclay[26] and operated in the ScanPyramids project that observed a big void in Khufu's Pyramid.[27] A telescope is composed of four identical detectors with an active area of $50 \times 50\,\mathrm{cm}^2$. A single detector is realized with a micro pattern technology called Micromegas that allows to read both the X–Y coordinates using readout strips. The pitch between the strips is about half mm and the total number of strips per detector, above 1,000. The strips are multiplexed using a patented algorithm so that the number of electronics channels needed to read the strips is reduced by one order of magnitude.

Figure 10. The Close Cathode Chamber (from Ref. 25).

This reduction implies a reduction of the cost and of the power consumption of the telescope. The high voltage, of about 2 kV, is provided by a dedicated high-voltage power supply card that works with 12 V of input voltage. The detectors operate with a flow below 0.5 liters per hour per telescope and the power consumption is about 35 W per telescope.

The use of muon radiography detectors in boreholes opens important perspectives in the field of civil engineering and mining. In order to cope with the geometrical constraints, detectors with a cylindrical geometry are being studied, with a combination of straight line and arc-shaped scintillator bars[28] to form 2-view detectors. One of the first prototypes of a cylindrical tracker, called MGR,[29, 30] was realized as early as in 2005 using 48 scintillating bars with wavelength shifting fibres to form the internal cylindrical surface, and two layers of scintillating fibres, 2 mm in diameter, disposed as counterclockwise helix and clockwise helix, respectively. The detector with diameter of just 14 cm allowed its use even in small diameter boreholes (see Fig. 11).

Figure 11.   The MGR detector inside a small diameter borehole (from Ref. 29).

The fibers were read by six multi-anode PMT for a total of 384 electronics channels. The total power consumption was 40 W, equally distributed between the front-end/readout electronics and the computer. The detector was tested in two Italian archaeological sites: Aquileia near Udine and the Traiano and Claudio port, near Rome.

References

1. L.W. Alvarez *et al.*, Search for hidden chambers in the pyramids, *Science*, **167**, 3919, 832–839, (1970).
2. H.K.M. Tanaka *et al.*, Cosmic-ray muon imaging of magma in a conduit: Degassing process of Satsuma-Iwojima Volcano, Japan, *Geophys. Res. Lett.*, **36**, L01304, (2009).
3. H.K.M. Tanaka *et al.*, Radiographic visualization of magma dynamics in an erupting volcano, *Nature Communications*, **5**, 3381, (2014).
4. W. Leo, *Techniques for Nuclear and Particle Physics Experiments*. Springer-Verlag, Berlin Heidelberg (1994).
5. G.F. Knoll, *Radiation Detection and Measurement*, 4th Edition. John Wiley & Sons Inc. (2010).
6. P.A. Zyla *et al.*, Review of particle physics, *Prog. Theor. Exp. Phys.*, **2020**, 8, 083C01, (2020).
7. S. Gundacker and A. Heering, The silicon photomultiplier: Fundamentals and applications of a modern solid-state photon detector, *Phys. Med. Biol.*, **65**, 17, 17TR01, (2020).
8. H.K.M. Tanaka *et al.*, Detecting a mass change inside a volcano by cosmic-ray muon radiography (muography): First results from measurements at Asama volcano, Japan, *Geophys. Res. Lett.*, **36**, L173602, (2009).
9. A. Anastasio *et al.*, The MU-RAY detector for muon radiography of volcanoes, *Nucl. Instrum. Meth. A*, **732**, 423–426, (2013).
10. F. Ambrosino *et al.*, The MU-RAY project: Detector technology and first data from Mt. Vesuvius, *JINST*, **9**, 02, C02029, (2014).
11. G. Saracino *et al.*, The MURAVES muon telescope: Technology and expected performances, *Ann. Geophys.*, **60**, 1, S01013, (2017).
12. M. D'Errico *et al.*, Muon radiography applied to volcanoes imaging: The MURAVES experiment at Mt. Vesuvius, *JINST*, **15**, 3, C03014, (2020).
13. N. Lesparre *et al.*, Density muon radiography of La Soufrìere of Guadeloupe volcano: Comparison with geological, electrical resistivity and gravity data, *Geophys. J. Int.*, **190**, 1008–1019, (2012).

14. J. Marteau *et al.*, Muons tomography applied to geosciences and volcanology, *Nucl. Instrum. Meth. A*, **695**, 23–28, (2012).
15. The ATLAS Muon Collaboration, The ATLAS Muon Spectrometer Technical Design Report, *CERN-LHCC/97-22 ISBN 92-9083-108-1* (1997).
16. The CMS Collaboration, The CMS experiment at the CERN LHC, *JINST*, **3**, S08004, (2008).
17. C. Carloganu *et al.*, Towards a muon radiography of the Puy de Dome, *Geosci. Instrum. Method. Data Syst.*, **2**, 1, 55–60, (2013).
18. M. Bedjidian *et al.*, Performance of Glass Resistive Plate Chambers for a high-granularity semi-digital calorimeter, *JINST*, **6**, P02001, (2011).
19. F. Ambrosino *et al.*, Joint measurement of the atmospheric muon flux through the Puy de Dome volcano with plastic scintillators and Resistive Plate Chambers detectors, *J. Geophys. Res. Solid Earth*, **120**, 11, 7290–7307, (2015).
20. M. Del Santo *et al.*, Looking inside volcanoes with the Imaging Atmospheric Cherenkov Telescopes, *Nucl. Instrum. Meth. A*, **876**, 111–114, (2017).
21. G. Saracino *et al.*, Imaging of underground cavities with cosmic-ray muons from observations at Mt. Echia (Naples), *Sci. Rep.*, **7**, 1181, (2017).
22. L. Cimmino *et al.*, 3D Muography for the search of hidden cavities, *Sci. Rep.*, **9**, 2974, (2019).
23. G. Baccani *et al.*, The MIMA project. Design, construction and performances of a compact hodoscope for muon radiography applications in the context of archaeology and geophysical prospections, *JINST*, **13**, 11, P11001, (2018).
24. K. Chaiwongkhot *et al.*, Development of a portable muography detector for infrastructure degradation investigation, *IEEE Trans. on Nucl. Sci.*, **65**, 8, 2316–2324, (2018).
25. G.G. Barnafoldi *et al.*, Portable cosmic muon telescope for environmental applications, *Nucl. Instrum. Meth. A*, **689**, 60–69, (2012).
26. S. Bouteille *et al.*, A Micromegas-based telescope for muon tomography: The WatTo experiment, *Nucl. Instrum. Meth. A*, **834**, 223–228, (2016).
27. K. Morishima *et al.*, Discovery of a big void in Khufu's Pyramid by observation of cosmic-ray muons, *Nature*, **552**, 386–390, (2017).

28. L. Consiglio *et al.*, Study of the light response of an arch-shaped scintillator with direct coupling to a Silicon Photomultiplier readout, *JINST*, **14**, P01014, (2019).
29. M. Basset *et al.*, MGR: An innovative, low-cost and compact cosmic-ray detector, *Nucl. Instrum. Meth. A*, **567**, 298–301, (2006).
30. M. Menichelli *et al.*, A scintillating fibres tracker detector for archaeological applications, *Nucl. Instrum. Meth. A*, **572**, 262–265, (2007).

Chapter 6

Three-Dimensional Muography and Image Reconstruction Using the Filtered Back-Projection Method

Shogo Nagahara and Seigo Miyamoto*

Earthquake Research Institute, The University of Tokyo
1-1-1 Yayoi, Bunkyo, Tokyo 113-0032, Japan
*miyamoto@eri.u-tokyo.ac.jp

In this chapter, we introduce a mathematical method using filtered back-projection for reconstructing the three-dimensional tomographic density image of a volcano, based on projection data obtained by multi-directional muography. The basic theories, some factors specific to multi-directional muography of volcanoes, and improvements to this method are described. An example of a simulation using real volcanic topography is also presented.

1. Introduction

X-ray radiography has evolved into X-ray computed tomography (CT), with the latter providing more information due to three-dimensional (3D) spatial resolution. Muography (i.e., muon radiography) can also provide 3D density images from observations in multiple directions. In this chapter, we describe a 3D density image reconstruction method based on multi-directional observational data. We also assess the performance of the method, mainly with respect to imaging volcanoes.

It is important to constrain the internal structure of volcanoes. For example, numerous studies have highlighted the importance of the shape of the volcanic conduit on the dynamics of eruptions. Koyaguchi and Suzuki[1] reported that crater and magma conduit shapes determine whether an eruption generates a pyroclastic flow or fall deposit.

Muography is a geophysical surveying method for imaging the internal density structure of an object using high-energy cosmic-ray muons. This enables imaging of the shallow density structure of volcanoes. For example, Tanaka et al.[2] applied muography to Satsuma–Iwojima, which is an active volcano located in southern Japan, and visualized vesiculated magma in the shallow conduit. This result confirmed the magma convection model that explains how volcanoes like Satsuma–Iwojima can efficiently release a large amount of volcanic gases.

One-directional muography provides only a density length, which is an integral of mass density times length along a path of muon. More than two observational directions are required to obtain spatial resolution along the muon path. For example, Tanaka et al.[3] attempted to reconstruct the three-dimensional density of the shallow crater at Asama volcano using muon observational data from two directions. Lesparre et al.[4] also conducted two-directional muography on La Soufriere of Guadeloupe volcano and compared the results with conventional geophysical surveys. Typically, two- or three-directional stereographic muography lacks the spatial resolution to resolve the detailed 3D structure of a volcano. However, recent technological developments have improved muon detectors (e.g., Refs. 5 and 6), which makes multi-directional muography increasingly practical.

It is also possible to acquire 3D spatial resolution by using other geophysical techniques. A joint inversion of one-directional muography and gravity observations revealed the 3D density structure of the Showa–Shinzan lava dome in northern Japan.[7,8] Rosas-Carbajal et al.[9] also conducted a joint inversion of three-directional muography and gravity measurements of the lava dome in Soufriere Hills volcano. These methods were based on inversion theory with an a priori covariance matrix and linear assumptions (e.g., Ref. 10). Such methods are also applicable in the case of multi-directional muography, but this chapter does not describe these in detail.

Like X-ray CT imaging, improved 3D spatial resolution can be obtained by observing a volcano from more directions. However, there are some differences: (1) The number of muon detectors around the volcano is limited due to the logistics and cost, (2) the muon intensity is much lower in muographic imaging, as secondary cosmic ray muons are generated by high-energy primary protons in the atmosphere at a constant flux, and (3) the cross-section length changes rapidly with elevation and, in the case of volcanoes, the muon beam has an elevation angle. Therefore, given the limited number and the total effective area of the muon detectors, it is necessary to optimize the placement of muon detectors and the angular resolution based on the volcano and/or features to be studied.

In the next section, we describe a 3D imaging method termed "filtered back-projection" (FBP) that uses two-dimensional angular images obtained by multi-directional muography, based mainly on the approximate reconstruction method of Feldkamp *et al.*[11] for a cone beam.

2. Filtered Back-Projection for Multi-Directional Muography

In this section, we provide an overview of FBP and its basic theory and describe the problems specific to muography and how these might be overcome.

Cosmic ray muons enter a volcano that has a specific internal density structure. How many muons reach a detector? This issue can be classified as a forward problem. We can, however, estimate the density structure inside a volcano from the attenuation of muons that pass through it. This is classified as an inverse problem. If the objective is to estimate the internal density structure by observations, the latter needs to be solved.

There are two main types of tomographic reconstruction methods, which are analytical and iterative. The former method is FBP (e.g., Ref. 12). We can further classify the latter into two categories, which are algebraic and statistical approaches. The algebraic methods are the algebraic reconstruction (ART; e.g., Ref. 13) and simultaneous reconstruction (SIRT; e.g., Ref. 14) techniques. The maximum likelihood–expectation maximization (ML–EM) technique (e.g., Ref. 15) is a statistical method.

The advantage of FBP is that it can produce relatively high-quality images with short computational times. However, it can be affected by artifacts that limit image quality. Another disadvantage of FBP is that it does not consider the statistics of the signal beams.

2.1. *Projection and back-projection*

In this subsection, we provide a basic introduction to projection and back-projection. For simplicity, let us first consider projection and back-projection with a parallel beam in 2D space.

As shown in Fig. 1, $\xi = x \cos \theta + y \sin \theta$ and $\eta = -x \sin \theta + y \cos \theta$. If $f(x, y)$ is the density distribution of an object, then the projection $p(\xi, \theta)$ is defined as follows:

$$p(\xi, \theta) = \int_{-\infty}^{\infty} d\eta \, f(x, y) \tag{1}$$

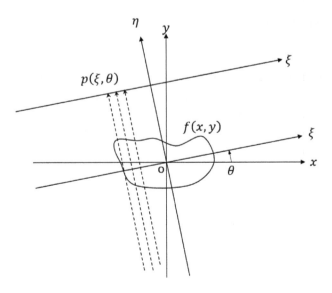

Figure 1.   A schematic representation of the projection and coordinate definitions.

If B is the operator of back-projection and $p(\xi, \theta)$ is the given projection data, the back-projection is defined as

$$B\{p(\xi, \theta)\} = \int_0^{\pi} d\theta \, p(\xi, \theta) \tag{2}$$

Note that $B\{p(\xi, \theta)\}$ is not equal to $f(x, y)$. The back-projection is a simple reconstruction technique, however, the reconstructed image tends to be blurred (Fig. 2).

2.2. *Filtered back-projection*

The filtered back-projection method is used to remove blur from an image. In the FBP method, the edges of the projection data $p(\xi, \theta)$ are emphasized by applying the following filters, before back-projection is conducted.

Let us assume that $P(k, \theta)$ and $F(u, v)$ are the Fourier transformations of the projection $p(\xi, \theta)$ and density distribution $f(x, y)$, respectively. According to the Projection–Slice Theorem, $P(k, \theta)$ is equivalent to $F(k \cos \theta, k \sin \theta)$, where $u = k \cos \theta$ and $v = k \sin \theta$. Using this theorem, the exact solution of $f(x, y)$ using projection data $p(\xi, \theta)$ is as follows:

$$f(x, y) = \int_{-\infty}^{\infty} dk \int_0^{\pi} d\theta \, |k| \, P(k, \theta) e^{i2\pi k(x \cos \theta + y \sin \theta)} \tag{3}$$

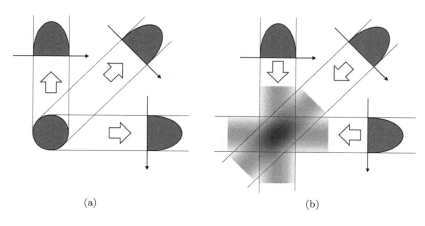

(a)   (b)

Figure 2. Schematic representations of the (a) projection and (b) back-projection.

By defining the new function $\hat{P}(k,\theta) = |k|P(k,\theta)$ and assuming that $\hat{p}(\xi,\theta)$ is the inverse Fourier transformation of $\hat{P}(k,\theta)$, the following equation holds:

$$f(x,y) = \int_0^\pi d\theta\, \hat{p}(\xi,\theta) \tag{4}$$

There is a similarity between Eqs. (2) and (4), but the filtered projection data $\hat{p}(\xi,\theta)$ are used instead of $p(\xi,\theta)$ in Eq. (4).

Ramachandran and Lakshminarayanan[16] proposed a method to treat Eq. (4) when the size element of the detector is significant. The Ram–Lak filter proposed by them is one of the most common filters in FBP. By defining the Nyquist frequency as $Q_{\max}$, the Ram–Lak filter $H(Q)$ in frequency space is

$$H(Q) = |Q| \quad (|Q| < Q_{\max})$$
$$H(Q) = 0 \quad\;\; (|Q| > Q_{\max}) \tag{5}$$

If the sampling interval of the projection is $\delta\xi$, the Nyquist frequency can be expressed as $Q_{\max} = 1/2\delta\xi$. The filter in real space can be expressed as

$$h(\xi) = \frac{1}{4\delta\xi^2}(n=0)$$
$$h(\xi) = -\frac{1-(-1)^n}{2\pi^2 n^2 \delta\xi^2}(n \neq 0) \tag{6}$$

where $\xi = n\Delta\xi$ (n is an integer). The image reconstruction $f(x,y)$ using the projection data $p(\xi,\theta)$ and filtering function $h(\xi)$ is written as

$$f(x,y) = \int_0^\pi \int_{-\infty}^\infty h(\xi - \xi')\, p(\xi',\theta)\, d\xi' d\theta \tag{7}$$

It is known that the Ram–Lak filter generates artificial noise when the number of directions is insufficient. An alternative filter was proposed by Shepp and Logan,[17] which smoothens the edges around $\pm Q_{\max}$ in Fig. 3(a) to decrease this artificial noise.

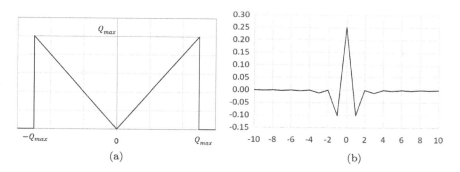

Figure 3. The shape of the Ram–Lak filter in frequency space and (b) real space where $\delta\xi = 1$.

2.3. *Three-dimensional imaging with the cone-beam approximation*

In this section, we introduce an approximation by Feldkamp *et al.*[11] when applying filtered back-projection to 3D density structures and cone beams. They proposed an approximation for constructing an FBP image of a defined cross-section using X-ray cone beams with an elevation angle in 3D space. This involves stacking of the reconstructed images of each cross-section shown in the previous subsection. This approximation is effective when the elevation angle is small.

If we define the coordinate system as shown in Fig. 4, then the equation for the reconstruction of Feldkamp *et al.*[11] is as follows:

$$\rho(x, y, z) = \frac{1}{2} \int_0^{2\pi} d\beta \int_{-X_M}^{X_M} dX \frac{D}{L_2^2\sqrt{1 + X^2 + Z_0^2}}$$
$$\times p(X, Z_0, \beta)h(X_0 - X) \qquad (8)$$

where $p(X, Z_0, \beta)$ is the density length, x, y, and z are the positions in 3D space, X and Z are the tangents of the azimuth and elevation angle, respectively, β is the observation point position at a counterclockwise angle with respect to the y-axis, and D is the distance between the observation point and origin. $L_2 = (D + x\sin\beta - y\cos\beta)$, $X_0 = (x\cos\beta + y\sin\beta)/L_2$, $Z_0 = z/L_2$, and $h(X)$ is the Ram–Lak

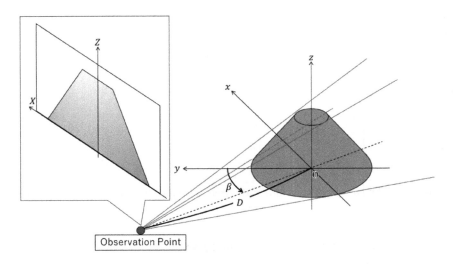

Figure 4. A schematic diagram of 1D muography and the definition of X, Z, β, and D in Eq. (8).

filter. For the function $h(X)$ in Eq. (8), we only need to replace the length variable ξ in Eq. (6) with X.

2.4. Application to multi-directional muographic imaging of a volcano

In this section, we describe how to apply FBP in multi-directional muographic imaging of a volcano. Multi-directional muography deals with significant changes in the length of cross-sections with elevation. The accuracy of the approximation in Eq. (8) worsens when there is a large change in path length along the vertical direction, as is the case for volcanoes. In order to improve the accuracy, it is useful to incorporate topographic information into Eq. (8) (e.g., Ref. 18). In many cases, the topography of a volcano (e.g., by aerial laser measurements) is generally available. If we define q and q_h (Fig. 5), the new approximation of the density length $p'(X, Z, \beta)$ can be written as

$$p'(X, Z, \beta) = \frac{q_h(X, z, \beta)}{q(X, Z_0, \beta)} p(X, Z, \beta) \tag{9}$$

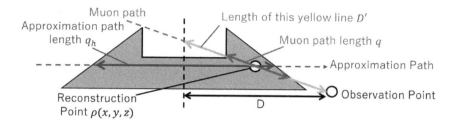

Figure 5. Schematic representation of the path length normalization approximation. In the approximation of Feldkamp *et al.*, the density length is $p' = (D/D')p$. In the approximation of Nagahara and Miyamoto,[18] the density length is $p' = (q_h/q)p$.

The final calculation formula is

$$\rho(x, y, z) = \frac{1}{2} \sum_{n=1}^{N} \delta\beta_n \sum_{m=1}^{M} \delta X_m \left(1 - \frac{X_m}{D(\beta_n)} \delta D_n \right)$$

$$\times \frac{D(\beta_n)}{L_2^2 \sqrt{1 + X_m^2}} \frac{p(X_m, Z_{0n}, \beta_n)}{q(X_m, Z_0 n, \beta_n)} q_h(X_m, z, \beta_n) h(X_0 - X_m)$$

(10)

where m and n are the indexes of X and β, respectively.

The beam intensity during muography is limited because the cosmic rays are environmental radiation. Therefore, it is necessary to consider the statistical error from the observed number of muons. However, FBP is unable to handle statistical errors. One solution to this issue was described in Nagahara and Miyamoto.[18] If i is the index of angular bins and N_i is the expected muon counts calculated by forward simulation, then the muon counts should follow a Poisson distribution, because the probability of a cosmic ray muon detection event is constant. We can evaluate how statistical errors affect the reconstructed images by adding random numbers into N_i according to a Poisson distribution, and by calculating the random error $\delta\rho^{\text{acc}}(x, y, z)$ as follows:

$$\delta\rho^{\text{acc}}(x, y, z) = \sqrt{\frac{1}{J-1} \sum_{j=1}^{J} \{\rho_j(x, y, z) - \rho(x, y, z)\}^2}$$ (11)

where j is the index of the trials, J is the total number of simulation trials, $\rho_j(x, y, z)$ is the reconstructed image of the jth trial, and $\rho(x, y, z)$ is the value calculated in Eq. (10) without a statistical error.

3. Performance Estimation with a Forward Modeling Simulation

In this section, we present an example of a simulation to determine the spatial and density resolution of a volcano by multi-directional muography and FBP.

3.1. *Models, assumptions, parameters, and procedures*

The volcano selected for this simulation is Omuroyama, which is located in Ito, Shizuoka Prefecture, Japan. Omuroyama volcano is a scoria cone that belongs to the Eastern Izu Monogenetic Volcanic Group. The volcano was active at 4 ka. Multi-directional muography was performed at this volcano because it has: (i) good accessibility for installation of muon detectors in all directions; (ii) no "shadows" from other objects; and (iii) an axisymmetric shape, although non-axisymmetric structures are expected to record differences in internal density.[19]

The topographic data used in this chapter were published by the Geospatial Information Authority of Japan. The minimum voxel unit of the volcanic body is a cube with dimensions of $20 \times 20 \times 20 \, \mathrm{m}^3$. The internal density structure has a checkerboard pattern, consisting of $100 \times 100 \times 100 \, \mathrm{m}^3$ with densities of 1.0 and 2.0 g/cm^3, respectively. The detectors were placed in a near-circular array around the volcano (Fig. 6).

The viewing range of each muon detector in the azimuth angle is ± 1.5 and the range of the elevation angle is 0.0–1.5 in terms of the tangent. For simplicity, the thickness of the detector was assumed to be negligible compared with the length of the effective sensitivity plane, and the muon detector efficiencies were all assumed to be 100% with no angular dependence. The model for muon flux was based on Honda *et al.*,[20] and the model for energy loss

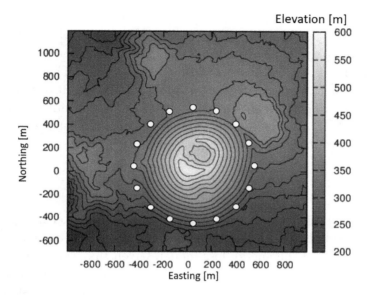

Figure 6. Topography around Omuroyama volcano and locations of the muon detectors in the simulation. White dots are the detector locations for the 16-directional muography. In the case of four-directional muography, the detectors were placed at the northern, eastern, southern, and western points. In the case of eight-directional muography, additional detectors were located in the northeast, southeast, southwest, and northwest. The small vent on the southern flank of the volcano is 450 m asl, the crater rim is at 535–580 m asl, and the crater floor is at 510 m asl.

of high-energy muons was based on Groom *et al.*[21] The effects of multiple Coulomb scattering were not considered in this simulation. We used two different total amounts of the product of the effective area × exposure time (i.e., 200 and 900 m^2d). For example, when the total amount is 900 m^2 d and there are 16 observation directions, the product of the effective area and observation period per direction is $900/16 = 56.25\,m^2$d. The statistical distribution of muon numbers observed in each angular bin of each detector was assumed to follow a Poisson distribution.

The simulation first calculated the density length along the muon directions from the topographic data, by taking into account the internal density structure and positions and orientations of each

180   S. Nagahara & S. Miyamoto

detector. Next, the critical energy at which a muon can pass through the density length was calculated for every angular bin, which determined the expected number of penetrating muons. The size of angular bins for this calculation was $(0.01)^2$. Then the muon numbers were merged into angular bins of $(0.05)^2$ or $(0.10)^2$ to ensure a minimum of 25 muon statistics. Let i, k be the index of angular bins, N_i^{merged}, N_k^{simu} be the number of muons in the angular bin after/before merging, the following equation holds:

$$N_i^{\text{merged}}(\psi_i) = \sum_k N_k^{\text{simu}}(\psi_i) \qquad (12)$$

where ψ_i is the average density in the angular bin after merging. We can uniquely determine the value of ψ_i from N_i^{merged} using the muon flux model and the energy loss model. Given that ψ_i is equal to the factor $p(X_m, Z_{0n}, \beta_n)/q(X_m, Z_{0n}, \beta_n)$ in Eq. (10), we can substitute this value into the final calculation of the 3D density reconstruction.

4. Results

Figure 7 shows the tomographic images of the reconstructed density of a horizontal cross-section at 470 m asl. The number of observation directions is 4, 8, 16, and 32, and the size of the angular bins is $(0.10)^2$ and $(0.05)^2$ in each 2D muographic image. Figure 8 also shows the computed tomography of a vertical cross-section at a northing of 150 m, under the same conditions as Fig. 7.

Figure 9 shows the density profile along an easting coordinate at 470 m asl and a northing of 150 m. The number of observation directions is 4, 8, 16, and 32. The total sum of the product of the effective area × exposure time is 200 and 900 m^2 d, with angular bins of $(0.10)^2$ and $(0.05)^2$, respectively. The difference between the center of the plots and the dashed lines indicates a systematic error of the FBP, while the error bar lengths are the random error $\delta\rho^{\text{acc}}(x, y, z)$ caused by the muon statistics. Figure 10 shows the density profile along the vertical line at an easting of -70 m and a northing of 150 m.

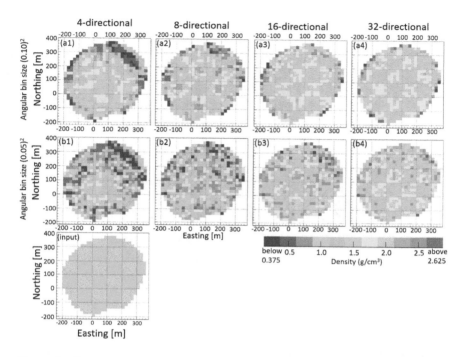

Figure 7. Reconstructed density profile for a horizontal cross-section at 470 m asl. (a1), (a2), (a3), and (a4) are the results of 4-, 8-, 16-, and 32-directional observations when the angular bin size of each muographic image is $(0.10)^2$. (b1), (b2), (b3), and (b4) are for an angular bin size of $(0.05)^2$. (Input) is the original input image.

5. Discussion

The difference between the input and reconstructed images becomes smaller as the number of observation directions increases (Figs. 7–10). Equation (4) is an exact solution at an arbitrary x, y. More contributions from various directions θ lead to a smaller difference between the input and reconstructed images. Given that increasing the number of observation directions is equivalent to obtaining various θ values for an arbitrary x, y, this leads to improved accuracy.

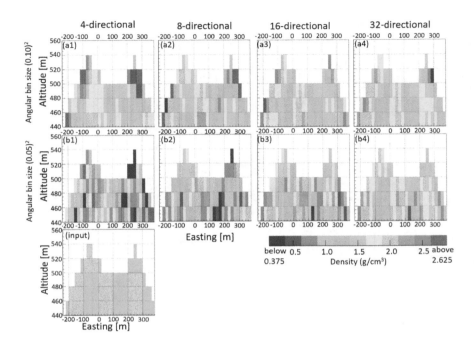

Figure 8. Reconstructed density profile for a vertical cross-section at a northing coordinate of 150 m. (a1), (a2), (a3), and (a4) are the results of 4-, 8-, 16-, and 32-directional observations with an angular bin size of $(0.10)^2$. (b1), (b2), (b3), and (b4) are for an angular bin size of $(0.05)^2$. (Input) is the original input image.

The same considerations can be applied to the bin size in 2D angular space. When comparing the $(0.10)^2$ and $(0.05)^2$ results, the latter better reproduces the input image. This is also due to the increase in the directions of θ in Eq. (4).

In Fig. 10, the random error caused by the muon statistics appears to depend on the elevation. This is because the number of penetrating muons increases as the thickness of the volcanic body decreases at higher elevations. This trend is also evident in Figs. 9(b) and 9(c) because the statistics of muon increase due to larger solid angle.

In Figs. 9 and 10, the random error caused by the muon statistics does not appear to depend on the number of observation directions. However, the random error varies with different-sized angular bins (Figs. 9(a)–9(b) and 10(a)–10(b)). This result differs from the statement by Nagahara and Miyamoto[18] that "this accidental error

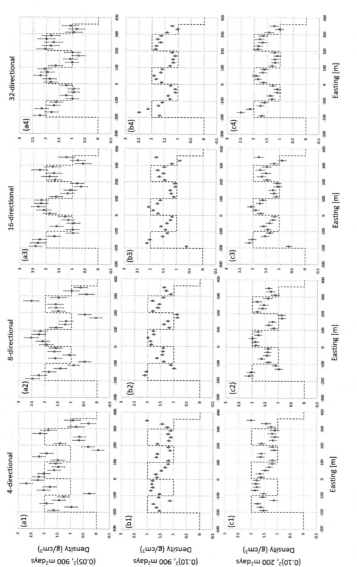

Figure 9. Horizontal density profiles at 470 m asl and a northing of 150 m. The horizontal axis represents the easting coordinate, and the vertical axis represents the density. Dashed lines are the input density, circles are the reconstructed density, and error bar lengths are the random errors due to the muon statistics $\delta\rho^{acc}(x, y, z)$ as defined in Eq. (11). (a1), (a2), (a3), and (a4) are the results of 4-, 8-, 16-, and 32-directional observations with an angular bin size for each muographic image of $(0.05)^2$ and a total sum of the product of the area and exposure time for each detector of 900 m^2 d. (b1), (b2), (b3), and (b4) are for the case of $(0.10)^2$ and 900 m^2 d. (c1), (c2), (c3), and (c4) are for the case of $(0.10)^2$ and 200 m^2 d.

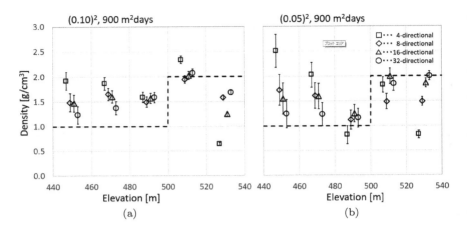

Figure 10. Vertical density profiles at an easting of $-70\,\mathrm{m}$ and a northing of $150\,\mathrm{m}$. The horizontal axis represents the elevation, and the vertical axis represents the density. Dashed lines are the input density, squares, diamonds, triangles, and circles represent 4-, 8-, 16-, and 32-directional data, respectively.

depends only on the total muon statistics for all observation points".
This statement is only correct when each muographic image has the
same angular bin size.

We should also consider the optimization of the angular bin size
for each muographic image. For example, in Fig. 9(a4), the random
errors are more significant than the difference between the input and
reconstructed density. However, the random error decreases as the
elevation decreases (Fig. 10). This implies that the size of the angular
bins should be optimized for the elevation angle of each muographic
image. For example, at a lower elevation, a larger angular bin size of
$(0.10)^2$ should be used.

6. Conclusions and Future Prospects

We have discussed the features specific to multi-directional muography, including the limited number of muon detectors, low signal
intensity, and rapid changes in cross-sectional length with elevation.
We have presented an example of a simulation to determine the
performance of the FBP method, which revealed systematic errors
in the reconstruction.

The recent developments in muon detector technology will allow multi-directional muography. For example, multi-directional muography was conducted around Omuroyama volcano (Refs. 22 and 23), which will provide new insights into this volcano and muon tomography.

However, few active volcanoes are like Omuroyama, where it is possible to locate muon detectors around the entire volcano. As such, most active volcanoes will have a biased orientation of instruments, which depends on the surrounding topography, ground surface, electricity, and logistics. Therefore, the systematic errors caused by this type of bias should be evaluated in future studies.

Acknowledgments

We thank Hideaki Aoki from Ike-kankou for collaboration on this study. We also thank Masato Koyama of Shizuoka University and Yusuke Suzuki of STORY Ltd. for discussions about Omuroyama volcano. This study was supported by JSPS KAKENHI Grant Numbers 19H01988 (Miyamoto), JSPS DC2 Fellowship JP19J13805 (Nagahara), and Izu Peninsula Geopark Academic Research Grant in 2018.

References

1. T. Koyaguchi, and Y.J. Suzuki, The condition of eruption column collapse: 1. A reference model based on analytical solutions, *Journal of Geophysical Research: Solid Earth,* **123**, 7461–7482, (2018). https://doi.org/10.1029/2017JB015308.
2. H.K.M. Tanaka, T. Uchida, M. Tanaka, H. Shinohara, and H. Taira, Cosmic-ray muon imaging of magma in a conduit: Degassing process of Satsuma–Iwojima Volcano, Japan, *Geophysical Research Letters,* **36**, L01304, (2009). https://doi:10.1029/2008GL036451.
3. H.K.M. Tanaka, H. Taira, T. Uchida, M. Tanaka, M. Takeo, T. Ohminato, Y. Aoki, R. Nishiyama, D. Shoji, and H. Tsujii, Three-dimensional computational axial tomography scan of a volcano with cosmic ray muon radiography, *Journal of Geophysical Research: Solid Earth,* **115**, B12332, (2010). https://doi:10.1029/2010JB007677.

4. N. Lesparre, D. Gibert, J. Marteau, J.-C. Komorowski, F. Nicollin, and O. Coutant, Density muon radiography of La Soufrière of Guadeloupe volcano: comparison with geological, electrical resistivity and gravity data, *Geophysical Journal International*, **190**, 2, 1008–1019, (2012). https://doi.org/10.1111/j.1365-246X.2012.05546.x.

5. K. Morishima, M. Kuno, A. Nishio, N. Kitagawa, Y. Manabe, M. Moto, F. Takasaki, H. Fujii, K. Satoh, H. Kodama, K. Hayashi, S. Odaka, S. Procureur, D. Attie, S. Bouteille, D. Calvet, C. Filosa, P. Magnier, I. Mandjavidze, M. Riallot, B. Marini, P. Gable, Y. Date, M. Sugiura, Y. Elshayeb, T. Elnady, M. Ezzy, E. Guerriero, V. Steiger, N. Serikoff, J. Mouret, B. Charles, H. Helal, and M. Tayoubi, Discovery of a big void in Khufu's Pyramid by observation of cosmic-ray muons, *Nature*, **552**, 386, (2017). https://doi:10.1038/nature24647.

6. L. Oláh, H.K.M. Tanaka, T. Ohminato, and D. Varga, High-definition and low-noise muography of the Sakurajima volcano with gaseous tracking detectors, *Scientific Reports*, **8**, 3207, (2018). https://doi:10.1038/s41598-018-21423-9.

7. R. Nishiyama, Y. Tanaka, S. Okubo, H. Oshima, H.K.M. Tanaka, and T. Maekawa, Integrated processing of muon radiography and gravity anomaly data toward the realization of high-resolution 3-D density structural analysis of volcanoes: Case study of Showa–Shinzan Lava Dome, Usu, Japan, *Journal of Geophysical Research Solid Earth*, **119**, 699–710, (2014). https://doi:10.1002/2013JB010234.

8. R. Nishiyama, S. Miyamoto, S. Okubo, H. Oshima, and T. Maekawa, 3D density modeling with gravity and muon-radiographic observations in Showa–Shinzan lava dome, Usu, Japan, *Pure and Applied Geophysics*, **174**, 1061–1070, (2017). https://doi:10.1007/s00024-016-1430-9.

9. M. Rosas-Carbajal, K. Jourde, J. Marteau, S. Deroussi, J.C. Komorowski, and D. Gibert, Three-dimensional density structure of La Soufrière de Guadeloupe lava dome from simultaneous muon radiographies and gravity data, *Geophysical Research Letters*, **44**, 6743–6751, (2017). https://doi:10.1002/2017GL074285.

10. A. Tarantola, Inverse problem theory and methods for model parameter estimation, *Philadelphia, Society for Industrial and Applied Mathematics* (2005). https://doi:10.1137/1.9780898717921.

11. L.A. Feldkamp, L.C. Davis, and J.W. Kress, Practical cone-beam algorithm, *The Journal of the Optical Society of America*, **A1**, 612–619, (1984).

12. S.R. Deans, *The Radon Transform and Some of Its Applications*, Dover Publications Mineola, New York, the united states of America. (2007).

13. R. Gordon, R. Bender, and G.T. Herman, Algebraic reconstruction techniques (ART) for three-dimensional electron microscopy and X-ray photography, *Journal of Theoretical Biology*, **29**, 471–481, (1970). https://doi.org/10.1016/0022-5193(70)90109-8.

14. A.H. Andersen, and A.C. Kak, Simultaneous algebraic reconstruction technique (SART): A superior implementation of the art algorithm, *Ultrasonic Imaging*, **6**, 81–94, (1984). https://doi.org/10.1016/0161-7346(84)90008-7.

15. K.-S. Chuang, M.-L. Jan, J. Wu, J.-C. Lu, S. Chen, C.-H. Hsu, and Y.-K. Fu, A maximum likelihood expectation maximization algorithm with thresholding, *Computerized Medical Imaging and Graphics*, **29**, 571–578, (2005). https://doi.org/10.1016/j.compmedimag.2005.04.003.

16. G.N. Ramachandran, and A.V. Lakshminarayanan, Three-dimensional Reconstruction from radiographs and electron micrographs, *Proceedings of the National Academy of Sciences of the United States of America*, **68**, 9, 2236–2240, (1971).

17. L.A. Shepp, and B.F. Logan, The Fourier reconstruction of a head section, *IEEE Transactions on Nuclear Science*, **21**, 21–43, (1974). https://doi:10.1109/TNS.1974.6499235.

18. S. Nagahara and S. Miyamoto, Feasibility of three-dimensional density tomography using dozens of muon radiographies and filtered back projection for volcanoes, *Geoscientific Instrumentation, Methods and Data Systems*, **7**, 307–316, (2018). https://doi:10.5194/gi-7-307-2018.

19. Y. Koyano, Y. Hayakawa, and H. Machida, The eruption of Omuroyama in the Higashi Izu Monogenetic Volcano Field, *Journal of Geography*, **105**, 4, 475–484, (1996). (In Japanese) https://doi.org/10.5026/jgeography.105.4_475.

20. M. Honda, T. Kajita, K. Kasahara, and S. Midorikawa, New calculation of the atmospheric neutrino flux in a three-dimensional scheme, *Physical Review D*, **70**, 043008, (2004). http://dx.doi.org/10.1103/PhysRevD.70.043008.

21. D.E. Groom, N.V. Mokhov, and S.I. Striganov, Muon stopping power and range tables 10 MeV–100 TeV, *Atomic Data and Nuclear Data Tables*, **78**, 183–356, (2001). https://doi.org/10.1006/adnd.2001.0861.

22. S. Miyamoto, S. Nagahara, K. Morishima, T. Nakano, M. Koyama, Y. Suzuki, A muographic study of a scoria cone from 11 directions using nuclear emulsion cloud chambers, Geoscientific Instrumentation,

Methods and 20 Data Systems, 11, 127–147, (2022). https://doi.org/1
0.5194/gi-11-127-2022.

23. S. Nagahara, S. Miyamoto, K. Morishima, T. Nakano, M. Koyama,
Y. Suzuki, Three-dimensional density tomography determined from
multi-directional muography of the Omuroyama scoria cone, Higashi–
Izu monogenetic volcano field, Japan, Bulletin of Volcanology volume
84, 94 (2022). https://doi.org/10.1007/s00445-022-01596-y.

Chapter 7

Muography and Geology: Volcanoes, Natural Caves, and Beyond

Jacques Marteau

IP2I, Univ Lyon, Univ Claude Bernard Lyon 1, CNRS/IN2P3, IP2I Lyon, UMR 5822, F-69622, Villeurbanne, France
j.marteau@ip2i.in2p3.fr

The present chapter describes applications of muography to some fields of geosciences: volcanology and geology, and opens on some of their "spin-offs" in industry and civil engineering. Despite the fact that muography first appeared in the context of archaeology, it has found a very prolific field of applications in the imaging of active domes, in Japan, France,and Italy for example. Beyond the strict production of images, in 2D or 3D, it has been used in joined analysis, e.g., with gravimetry and seismic data, to probe the inner structures and dynamics of the active domes. In parallel, the improvements of muon detectors' sensitivity make it possible to study the general features — density and volumes — of geological layers from underground measurements, by essence limited in statistics.

1. Introduction

From the early investigations of L. Alvarez performed in the Egyptian Chephren of Alvarez (1970)[1] to the recent results of the ScanPyramids project,[a] muons have gained in popularity not only in the archaeology domain but also in geosciences in general and even nowadays among industrial manufacturers. Muons are elementary particles and usually belong to the "high energy physics" (HEP) scientific world, one of the largest experiments ever built, like the

---

[a]http://www.scanpyramids.org/.

LHC at CERN[b] in Switzerland — the world's largest accelerators complex — or Super-Kamiokande in Japan[c] — the world's largest underground detector. The world where those particles are at the same time the subjects of research and the tools to achieve the results.

A matter of size and technology: reduced and autonomous particle detectors have been designed and made available outside the HEP world. And innovative techniques have developed which use elementary particles not for themselves but as tools for interdisciplinary sciences. This is the case for muography where different technologies are nowadays used, from the early "spark chambers" of Alvarez to "nuclear emulsions", "plastic scintillators", "resistive plate chambers," and "micromegas".

Muon imaging has emerged as a powerful method to complement standard tools in Earth Sciences. The general features of this technique are nowadays quite popular and they basically rely on the detection of atmospheric flux, where muons represent the largest proportion of charged particles reaching the surface of the Earth. They are secondary products of primary cosmic rays (high-energy protons and light nuclei) interacting with the atmosphere. The basic cascades leading to the muons' production are, without going into details: primary cosmic rays + oxygen/nitrogen nuclei → parent mesons (pions/kaons) → secondary muons. The intermediate stage of this production chain requires the decay of the so-called "parents", pions and kaons. For general properties of particles, we refer to the Particle Data Group reviews.[d] These decays require that there are no re-interactions of the parents with the environment. Since this mean free path depends on the particle's energy and on the atmosphere density, one sees that the muons' flux and spectrum may be affected by the environmental conditions. It is worth mentioning that part of the incident muons are also absorbed on their way down to Earth's surface with a probability depending as well on the atmosphere's density.

---

[b]https://home.cern/fr.
[c]http://www-sk.icrr.u-tokyo.ac.jp/sk/index-e.html.
[d]https://pdg.lbl.gov/index.html.

The rather low interaction cross-section of muons with matter ensures that they may reach the Earth's ground level, but furthermore that they may significantly penetrate large and dense structures that one would like to scan. At the same time, their sensitivity to the atmospheric conditions in which they are produced may provide valuable information on some not-so-easily accessible parameters such as, e.g., the temperature at the top of the stratosphere. Therefore, the range of applications of muography may be very large, one of the main reasons of today's domain expansion. The chapter is organized as follows. After a section devoted to a reminder of muography basics, two sections detail recent results in volcanology (Section 3) and natural caves imaging (Section 4), giving examples of geotechnical applications of interest for non-invasive controls in industry and civil engineering.

2. Muography Basics: Direct and Inverse Problem

2.1. *Detecting muons: The direct problem*

The starting point of muography is of course the detection of the muons, the "direct problem". This detection is quite simple since atmospheric muons are charged leptons. For a general review on muons' properties, see the Particle Data Group data compilation.[e] As they cross matter, they will interact with the charges (electrons and nuclei) of the medium. This will result in a loss of a fraction of their total energy and also in a deviation of their trajectory. These properties, sensitive to the density of the target (and to a lower extent to its Z/A ratio between the number of protons and the total number of nucleons), are exploited in the two different modes of muography called "absorption muography" and "scattering muography".

Atmospheric muons at ground level have an average energy usually ranging close to the minimum of ionization (see Ref. 2). For a more general review on the properties of particles through matter, see the Particle Data Group.[f] This energy loss may be

---

[e]https://pdg.lbl.gov/2020/tables/rpp2020-sum-leptons.pdf.
[f]https://pdg.lbl.gov/2020/reviews/rpp2020-rev-passage-particles-matter.pdf.

converted into various types of signals which constitute the basic "hits": charge avalanches in gaseous detectors such as Resistive Plate Chambers (RPC) or Micro-Megas, silver atoms in nuclear photographic emulsions or photons in scintillation detectors[3] The hits left by the muons in the detectors and collected together to build their trajectory, the muography detectors, belong to the "trackers" category.

The absorption mode is the same as for the X-ray medical imaging. One infers the mass distributions inside a given target from the measurement of the reduced muons flux due to their interaction with the matter of the target. The scattering mode allows the reconstruction of the mass distributions from the measurement of the muons' trajectory deviation angles upstream and downstream the target. It is usually restricted to small targets while the absorption mode is well-suited for large volume imaging.

The minimal requirements on the detection devices used for muography are therefore the tracking capabilities, whose performance is measured in terms of spatial (and/or angular) resolution, usually driven by the size of the detectors' segmentation or pixels, and in terms of timing resolution. The standard configuration for a tracker is to have parallel detection planes (more than 3) with XY resolution. This allows the minimal track reconstruction in the event-building procedure. Good spatial and angular resolutions are absolutely necessary if one wants to precisely scan details in rather small objects. These requirements are less stringent for large structures, such as volcanic domes where the measured fluxes are strongly reduced because of the target's opacity. For such measurements, the important parameter is the detector's acceptance, i.e., its capability of collecting the maximal number of muons for a given active surface as described in Sullivan (1971).[4] Larger matrices offer a larger detection area which reduces the acquisition time for a given angular resolution as detailed in Lesparre *et al.* (2010).[5] Good timing performances are also required for background reduction and time-of-flight measurements. The background rejection is important for outdoor applications where one needs to eliminate random coincidences and requires fine timestamps, of the order of the nanosecond or below. As its name implies, time-of-flight measurement consists in measuring the time taken by the muon to cross the detector and discriminating whether it was propagating downwards or upwards. This is absolutely vital if one wants to assess whether the muon crossed the target or if it belongs to the background. In this case a sub-nanosecond resolution is required for meter-scale detectors.

Figure 1 illustrates two implementations of the direct problem where the muon detector (pictures of the left column) is located either on the slope of an active dome (the Soufrière of Guadeloupe,

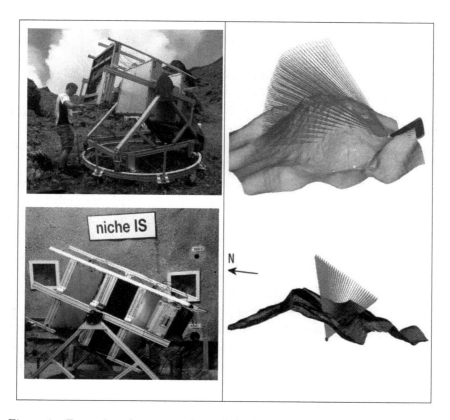

Figure 1. Examples of muon trackers of the Diaphane project (left) and sketch of the muon trajectories falling into their acceptance (right). Top. Open-air installation on the slope of the Soufrière of Guadeloupe ("Rocher Fendu" position). Bottom. Underground location in the Mont-Terri laboratory.

Lesser Antilles, France) or in a gallery of the underground Mont-Terri laboratory (Jura, Switzerland). The trackers use plastic scintillators as detection medium and have been operated within the Diaphane project.[g] The sketches in the right column represent all muons' trajectories falling into place by the detector's acceptance.

2.2. *From data to images: The inverse problem*

The most difficult step in muography is the so-called "inverse problem", when going from raw data to reconstructed mass distributions. For a general presentation of these general features, see Nagamine (2003).[6] Indeed muography is not suited for standard imaging techniques, like a Radon transform widely used in medical imaging for example. One of the main reasons is the limited statistics, imposed by the natural atmospheric muons' flux.

On top of the statistical limitation, there are intrinsic ambiguities for single-point measurement in absorption mode since the detector measures the attenuation of the muon flux integrated all over the path of the muons inside the studied target of density, i.e., its "opacity" defined as :

It is clear from this definition that a single measurement of a muon deficit (negative anomaly) or a muon excess (positive anomaly) w.r.t. a given model leads to an infinite number of possibilities as to the precise location of this anomaly along the path.

Going from an opacity map to a density map requires therefore a model or more generally an "inversion technique" that provides the most probable mass distribution functions inside the target (for a review, see Ref. 5). The inverse problem needs to be constrained by the available "a priori" information (in the Bayesian language) but also driven by the data quality. And this imposes the requirements on the detector performance in terms of acceptance, resolution, stability in operation, duty cycle, etc.

Typical apparent density is displayed in Fig. 2 for the case of the Soufrière of Guadeloupe seen from the position of Fig. 1. The

---

[g]https://diaphane-muons.com/.

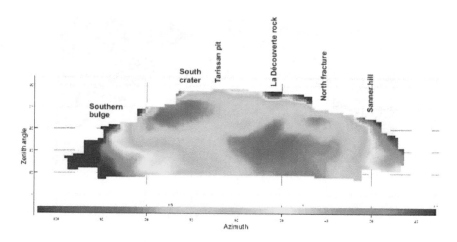

Figure 2. Apparent density map of the dome measured from the East side of the Soufrière of Guadeloupe volcano. The blue regions correspond to negative density anomalies while the red regions correspond to positive density anomalies.

structural image exhibits a highly heterogeneous structure for this dome, from the low-density regions (in blue on the map) in the center and below the South-East crater (the most active zone of the volcano, where vents are recorded at high velocity) to the high-density regions close to the edges of the dome mainly constituted by basaltic rocks.

3. From Volcanoes to Geotechnical Imaging

On volcanology, there are well-established collaborations for muon imaging of volcanoes, which led to huge progress in the understanding of the internal structure and magma dynamics. The obvious advantage of muography is the possibility to perform target scanning from remote positions and in an autonomous way, without on-site operators as for, e.g., ERT measurements. The technique is also well-suited for continuous 24/7 measurements essential for monitoring purposes. Given the dangerousness of many active domes, those features make muography a perfect tool for volcanological surveillance.

J. Marteau

Figure 3. Muography of Showa-Shinzan lava dome (a) compared to its photography (b).

Recent progress in methodological developments are impressive and have been obtained mainly in volcanological applications. A non-exhaustive list of the most visible achievements is as follows:

- Structural imaging of a volcanic dome from one or more different points of view: Idowake, Ontake, Showa-Shinzan, etc in Japan, Stromboli, Etna, Vesuvius, etc. in Italy, Puy-de-Dôme, Soufrière of Guadeloupe in France, Mayon in the Philippines... An example is given in the following Fig. 3.[7]
- 3D density structure of a lava dome[8] in complete coincidence with the 3D structure obtained with ERT techniques.[9]
- Identification of huge mass and energy transfers associated with the hydrothermal activity.[10, 11]
- Simulation and rejection of the upward muons flux effects.[12]

- Simulation and subtraction of the muon diffusion at the surface of the volcano effects.[13]
- Joined muon-gravimetry inversion to perform 3D reconstruction of the dome.[14]
- Combined measurements of muography and seismic noise data.[11]

In the following, we focus on particular projects chosen among 3 major collaborations in Japan, France, and Italy.

3.1. *Pioneering works in Japan*

Early applications of muography applied to volcanology have been led by the ERI group of Pr. Tanaka using different types of detectors: nuclear emulsions and scintillators (see Tanaka (2007, 2008, 2009)[7, 15, 16] and references therein). For instance, Mt. Asama was the first volcano which was imaged with muography in 2006[7] using so-called Emulsion Cloud Chambers (ECC) developed in the scope of the neutrino oscillations search experiments, e.g., OPERA.[17] After its eruption in 2004, Mt. Asama could not be accessed, making impossible the use of conventional geophysical methods, such as electromagnetic and seismic techniques. The ECC's detection area was $0.4\,m^2$ and it was located inside a 1 meter deep vault at 1 km of the crater's summit, giving an overall 10 meter image resolution. The data analysis requires emulsion films development and readout with a microscope to build the tracks.[h] The density distribution of the summit crater was extracted by GEANT4 simulation including a topographic map of Mt. Asama. The resulting image is shown in Fig. 4.

The second example of these pioneering experiments was performed during an eruption period of the Satsuma–Iwojima volcano in 2013. The interest here is the observation of the dynamics of the eruption in real-time. Of course, ECC techniques cannot be applied for real-time measurements since this is a passive method. Therefore,

---

[h]Details of the emulsion techniques are given in a separate chapter of the book.

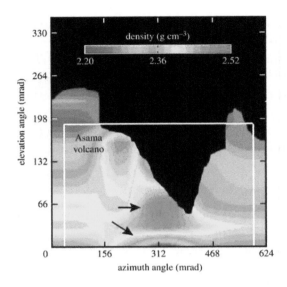

Figure 4. First muographic image of a volcano: Mount Asama, Japan. The reconstructed average density distribution of the summit crater shows the solidified magma (red region) at the crater floor.

a scintillator-based tracking system was developed, with six $2\,\mathrm{m}^2$ scintillator planes with a 10 cm position resolution. The tracking system was deployed at 1.4 km from the summit crater. Five 10 cm thick lead absorbers were installed between the scintillators to absorb the low-energy particles and reject this background. The difference between an empty and a filled crater with 2σ (95%) confidence level was observed after 3 days of data-taking. Figure 5 shows the first time-sequential muon radiographic animation about the magma dynamics inside an erupting volcano with the 1σ (68%) confidence level upper limit of the average density plotted along the muon path. During the eruptions, from June 14 to 16, and from June 29 to July 1, the higher dense magma ascended in the crater, as shown in the upper left and in the middle right panels of Fig. 5. The observed dynamics of the volcanic gases and the magma body was found consistent with the models.

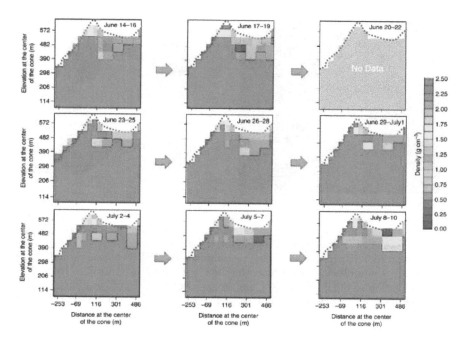

Figure 5. Muography of the magma column dynamics during the eruption of Satsuma-Iwojima volcano.

3.2. *Diaphane and the Soufrière of Guadeloupe*

Diaphane is a french interdisciplinary collaboration between particle physicists of the CNRS-IN2P3 institute (IP2I Lyon) and geophysicists of the CNRS-INSU institute (Géosciences Rennes, IPG Paris). It involves also the related universities.

These collaborators build several scintillator-based trackers on the concept of distributed smart sensors, adapted to harsh environmental conditions in totally autonomous modes and used them to make continuous muon tomography experiments for more than 10 years on the Soufrière of Guadeloupe volcano, in the Lesser Antilles, making this volcano the most equipped in the world with a network of 6 telescopes simultaneously taking data in the 2017–2020 period (Figs. 2 and 6).

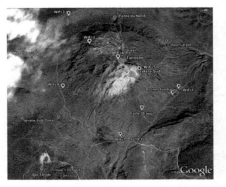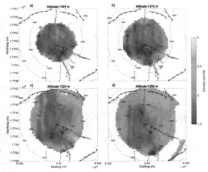

Figure 6. Left. Summary of the experiments on the Soufrière of Guadeloupe with 6 muon detectors operated around the dome (blue diamonds). The red triangle represents the location of the CG3 gravimeter for monitoring. The yellow symbols represent the WiFi relays necessary to establish the link between the telescope and the volcano observatory where the data are stored in real-time. Right. 3D reconstruction of the Soufrière of Guadeloupe's dome: slices of constant altitude. The results have been obtained by coupling muography data and gravimetry data in a process of joined inversion for the density.

The collaborators were the first to perform a 3D joined inversion of muography and gravimetry data (Fig. 6), in perfect agreement with the 3D inversion model of the dome from ERT measurements.[9] The joined inversion in that particular case is straightforward since both methods measure the same observable density. Technically, one has to invert the following problem:

where is the forward kernel, the data vectors, subscripts and denote the gravity and muon case, respectively, is the density distribution, and accounts for a possible density offset between the gravity- and muon-inferred models.

The Diaphane collaboration obtained a less obvious and promising result by combining muography and seismic data from geophones distributed at the summit of the Soufrière.[11] They found the appearance of an active hydrothermal focus located 50–100 m below the summit of the volcano, by observing the coincidence between a 2-day sequence of oscillations of both amplitude and dominant frequency of the seismic noise followed by a sharp decrease of the

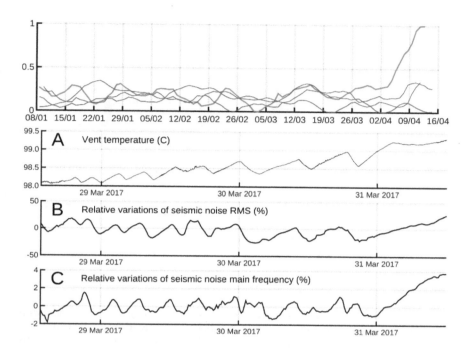

Figure 7. Top. Time variations of the muon flux across different domains of the lava dome. The red curve is for the bundle of lines of sight covering the seismic source zone. The blue curves are for adjacent areas. Oscillation amplitudes are arbitrarily set to a common value. Bottom. Time variations of vent temperature (A), seismic noise RMS (B), and dominant frequency (C).

bulk density visible as an augmentation of the muon flux crossing the active zone. This is illustrated in Fig. 7 where the sequence of seismic and muon events is displayed.

A spectral coherency analysis of the seismic noise followed by a bandpass filtering allowed to localize the seismic source by an unconstrained back-propagation algorithm, which pointed toward a rather small volume of about $10^4 \, \mathrm{m}^3$. This convergence zone falls in the sub-acceptance of the muon detector which "sees" the sharp flux increase, in coherence with the seismic activity (Fig. 8).

Besides their pure scientific interest, these results are of much importance from a volcanological point of view to constrain flank

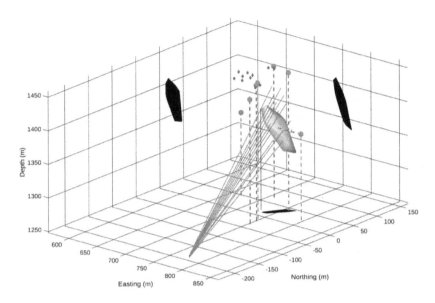

Figure 8. Location of the seismic noise source volume. The yellow body represents the 3D convergence zone of the wavefronts back-propagated from the geophones located on top of the lava dome (blue dots). The red dots represent the presently main active vents. The black patches are the projections of the source volume onto the faces of the 3D block diagram. The fan-like bundle of straight lines represents the lines of sight of the telescope measuring the flux increase in Fig. 7.

destabilization models and risk assessment[18, 19] and represent an interesting perspective for a time analysis of the volcanological dynamics.

3.3. MURAVES (Vesuvius) and TOMUVOL (Puy-de-Dôme)

In this section, we present two recent results obtained with two original techniques: plastic scintillators with a triangular shape and a SiPM readout for the MURAVES collaboration and Glass-Resistive Plate Chambers (GRPCs) for the TOMUVOL one.

MURAVES is a collaboration, between INGV, INFN, UCL, and UGent, which runs three detectors located on the Vesuvius slopes in Italy, one of the most dangerous volcanoes in Europe.[20] This project was initiated long ago with the MURAY project of collaborative

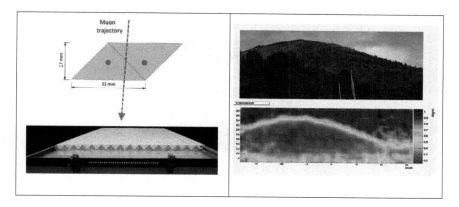

Figure 9. Left. MURAVES scintillators, schematics, and detection plane. Right. MURAVES structural imaging of the Vesuvius.

efforts around methodological and instrumentation developments.[21] An interesting feature is the use of triangular-shaped scintillator bars, obtained by extrusion in the Fermilab laboratory. The combination of information from two adjacent bars allows increasing the spatial resolution by weighting the amount of light collected (Fig. 9).

TOMUVOL[i] (Tomographie MUonique des VOLcans) is an interdisciplinary collaboration initiated in 2009, joining particle physicists and volcanologists from three French laboratories: LMV and LPC, located in Clermont-Ferrand, and the IP2I Lyon. The project benefits from a reference site located near Clermont-Ferrand: the Puy de Dôme (alt. 1,464 m a.s.l.), an extinct 11,000-year-old volcanic dome in the Massif Central, south-central France. It has a remarkable structure with two domes originating from two subsequent eruptions, which occurred within a short time interval. Its density structure is therefore expected to be complex with large variations. This site has interesting benchmarking features for methodological developments in geosciences imaging.

---

[i]https://wwwobs.univ-bpclermont.fr/tomuvol.

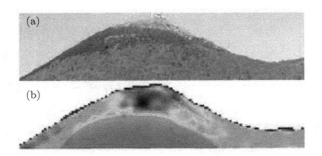

Figure 10. Map of the scaled transmission through the Puy de Dôme as seen over 7 months from the Grotte de la Taillerie with a $\sim 1/6\,\mathrm{m}^2$ detector.

The TOMUVOL muon tracker is made of parallel planes of GRPCs, operated in avalanche mode. This technology allows a high segmentation, in $1\,\mathrm{cm}^2$ cells. GRPCs are rather cheap, robust, highly efficient ($\sim 95\,\%$), with a detection rate up to $100\,\mathrm{Hz}$ and a low noise level, less than $1\,\mathrm{Hz}\cdot\mathrm{cm}^{-2}$. A typical mass distribution map obtained during the first data-taking campaign is displayed in Fig. 10.

The previous review is not exhaustive of course, and there are other collaborations covering other volcanoes : Etna,[22, 23] Stromboli,[24] just to quote Italian volcanoes. But it gives an overview of methods and detection techniques used to perform structural and dynamical imaging of volcanic domes. Those methodological developments and successful results have attracted the industrial sector's attention for non-invasive/non-destructive controls of large and/or inaccessible structures like, e.g., blast furnaces, nuclear power plants, storage silos, etc. In the following section, we give an important example of muography applied to the search of the nuclear reactor's remnants in Fukushima Daiichi, after the 2011 tsunami.

3.4. *Nuclear reactor investigation*

The nuclear power reactors of Fukushima Daiichi were heavily damaged by the giant earthquake and subsequent tsunami that occurred in March 2011. Before decommissioning it is necessary to acquire information about the status of the reactors. Because of the high radiation inside and around the reactor buildings, it was decided to use muography techniques to study the situation of the reactor, specifically the status of the nuclear fuel assemblies.[25] A scintillator-based detector system has been built and calibrated at the nuclear reactor of the JAPC at Tokai, Ibaraki, Japan. This resulted in a successful imaging of the inner structure of the reactor.

The detector was then brought at the Unit-1 reactor of Fukushima Daiichi and operated there from February 2014 to June 2015. The muon tracking detectors were placed outside the reactor building. They consisted of three XY sets of 1 cm-wide plastic scintillation bar counters arranged in planes of 1 m × 1 m. The muon telescope was housed in a container of 10 cm-thick iron to suppress the effects of the environmental radiation (as high as 0.5 mSv/h in the area).

The important result of this study is the success in identifying the inner structure of the reactor complex, such as the reactor containment vessel, pressure vessel, and other structures of the reactor building, through the concrete wall of the reactor building (Fig. 11). It was found that a large number of fuel assemblies were missing in the original fuel loading zone inside the pressure vessel. The natural interpretation is that most of the nuclear fuel was melted and dropped down to the bottom of the pressure vessel or even

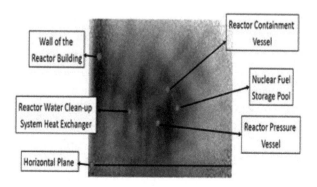

Figure 11. Image of the Unit-1 reactor of Fukushima Daiichi after 90 days of observation by the detection system placed at the north-western corner of the reactor building.

below. This important application is a very promising result opening perspectives in the dismantling activities of nuclear power plants, by putting constraints on the civil engineering structures and/or nuclear reactors in order to assess the risks and prevent hazards.

4. From Natural Caves to Civil Engineering

Muography may be applied in different contexts such as the characterization of geological layers from underground measurement locations. Of course, the muons flux decreases with depth, but the background is also less present, leaving room for a large variety of studies with important applications for civil engineering, prospecting, and mining. We detail an example of such investigations in a research infrastructure, devoted to methods developments in geosciences, the Mont Terri underground laboratory in Switzerland.[j]

4.1. The underground Mont Terri laboratory: Geological muography

The Mont Terri Project is an international research project for the hydro-geological, geochemical, and geotechnical characterization of a clay formation (Opalinus Clay). It is operated by Swisstopo.[k] It has many instrumented galleries where one can in particular install muon detectors at various places (Figs. 1 and 12).

In this laboratory, first trials were performed to characterize the opalinus clay layer, separate geological layers of close density with a joined muon-gravimetry measurement,[14] monitor the hydrothermal activity of a karstic system. The possible applications of those studies range from the R&D on the long-term storage of nuclear waste or on the burial of CO_2, to hydro-geology, characterization of karstic systems, and studies on the physics of the atmosphere as a by-product of muography measurements. Indeed, since the overburden acts as a "high-energy-pass filter" it is possible to study the variation in time of

---

[j]https://www.mont-terri.ch/fr/page-d-accueil.html.
[k]https://www.swisstopo.admin.ch/en/home.html.

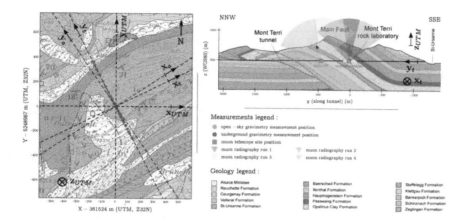

Figure 12. Left. Mont Terri anticline geological map. The green and blue dots correspond to the gravimetry measurement positions (respectively, open-sky and underground) and the two red squares are the sites occupied by the muon telescope. Right. Side view of the Mont Terri anticline. The colored cones taking their apex on the underground niches show the areas scanned by the muon telescope during the different acquisitions.

the high-energy muons flux and its correlation with the temperature variations at the top of the atmosphere.

Various data-taking campaigns took place at different locations and with different objectives (Fig. 12):

- 2009–2011: IS, SHGN, PP, EB niches.
- 2012–2015: PP niche and QQ niche (muon-gravity coupling).
- 2016–2017: IS niche (hydro-geological monitoring).

The characterization of the geological layers has been performed from four different acquisitions, as indicated in Fig. 12. The results, expressed in terms of measured muons flux compared to the theoretical flux, along the anticline axis, that would be detected if the Mont Terri anticline was homogeneous and for various apparent densities are displayed in Fig. 13. We can notice how the measured flux is well anti-correlated with the apparent thickness∼: the thicker the overburden, the weaker the muon flux. We also see how different runs from 2 locations recover coherently when they look at the same

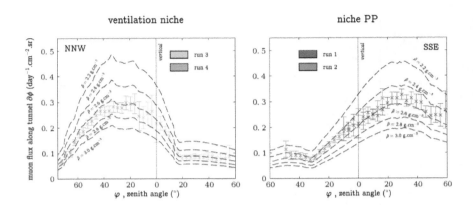

Figure 13. The colored crosses with error bars correspond to the measured muons absolute flux as a function of the zenith angle and along the Mont Terri tunnel direction. The error bars are plotted with a 95% confidence interval. The black dotted lines correspond to the theoretical flux for various fluxes' average densities specified directly on the curves.

zones, a sign that the detector's performances are well under control. It is worth noticing also that for the apparent density is decreasing to, signaling the existence of an area largely altered by the rain water.[26]

Those results are also expressed in terms of a joined muon-gravimetry analysis with a parametrization of the geological layers frontiers as Bezier curves. The comparison of muon-only, gravimetry-only, and muon-gravimetry analysis show how the constraints brought by the muography above the horizon manage to improve the resolution of gravimetry in the whole phase space.[8, 14]

4.2. *Observation of Sudden Stratospheric Warmings event*

The state-of-the-art of the monitoring technique has been assessed on known structures like water tower tanks where the water level is controlled by standard gauges and is used to cross-check the registered changes.[27] The corrections of the fluxes with the atmospheric parameters have been parameterized on the form of a first-order expression:

where the 2 parameters are adjusted for on the data themselves. The sensitivity of the muons to those corrections is energy-dependent. Pressure corrections are more important for the lowest energy muons since the changes in the atmospheric density affect directly the muons mean free path. Temperature changes at the top of the atmosphere modify the mean path of the muons parents, the short-lived pions and/or kaons. Since the lifetime of those particles increases with their energy due to the Lorenz boost, the high-energy part of the spectrum will be more efficiently suppressed at low temperature, that is for a larger atmosphere density.

Those effects have been measured by various experiments (notably underground neutrino experiments) with different overburdens, acting as a (high-pass) filter for the muons energy. They have been correlated with well-known atmospheric phenomena such as Sudden Stratospheric Warmings (SSW).[28]

4.3. *Tunnel-boring machines*

Civil engineering may be a rich field of applications for muography.[29,30] Here we focus on tunnel-boring projects which require a considerable amount of planning, well ahead of drilling operations. Despite these precautionary measures, unexpected ground geological features (local density variations, cavities, unstable superficial ground, instabilities induced by the driller) impose real-time adaptations, possibly expensive, of the drilling operations.

The direct relation of muon flux absorption with the density of a given medium makes muography a promising solution to provide a real-time density analysis of geological objects in front of the tunnel-boring machine (TBM). A picture of such TBM is displayed on Fig. 14.

To test its applicability during the drilling of the "Grand Paris Express" subway network, a scintillator-based muon detector was used inside the TBM. Because the detector moves forward, the muon flux crossing a particular geological object measured by the telescope has different directions with respect to time, allowing 3D density estimates. An example of apparent density 3D reconstruction

Figure 14.　Left. Picture of a Tunnel-boring Machine (TBM) of the Herrenknecht company. Right. Artistic view of a TBM while digging an urban tunnel (right).

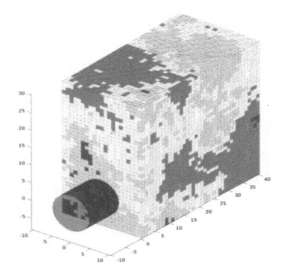

Figure 15.　Online 3D apparent density reconstruction averaging the last 50 meters of the TBM movement (credits J. Marteau, IP2I).[31,32] Each cube is $1\,m^3$ volume.

distribution of the ground, using inhomogeneous Poisson likelihood, is presented in Fig. 15. The algorithms for this particular reconstruction removing systematic noise from the buildings, caves, etc are patented, but they represent a powerful tool for civil engineering exploration methods with moving detection devices.

5. Conclusions

This chapter presented general features of muography applied to volcanology and characterization of geological layers. With different detection techniques, many collaborations around the world succeeded to scan the inner part of active volcano domes providing unprecedented inputs to volcanological models. The real-time measurements capabilities of the muography techniques open the field of monitoring, so powerful in the context of risk mitigation and hazard management.

The success of the method opens the door to a large number of applications, from Earth Sciences to non-invasive and non-destructive controls in the industry. With the technological improvements, the detectors' compactifications, and miniaturization, a lot of yet unexplored domains will be explored, either with underground measurements or with open-air measurements, the two main operation modes that we have illustrated with examples in geology and volcanology.

References

1. L.W., Alvarez, J.A. Anderson, F.E. Bedwei, J. Burkhard, A. Fakhry, A. Girgis, A. Goneid, F. Hassan, D. Iverson, G. Lynch, Z. Miligy, A.H. Mousaa, M. Sharkawi, and L. Yazolinio, Search for hidden chambers in the pyramids, *Science*, **167**, 832839, (1970).
2. T.K. Gaisser, R. Engel, and E. Resconi, *Cosmic Rays and Particle Physics*, Cambridge University Press (2016).
3. W.R. Leo, *Techniques for Nuclear and Particle Physics Experiments: A How to Approach*, Springer (1987), p. 368. ISBN: 9783540572800.
4. J.D. Sullivan, Geometrical factor and directional response of single and multi-element particle telescopes, *Nuclear Instruments and Methods in Physics Research A*, **95**, 5–11, (1971).
5. N. Lesparre, D. Gibert, J. Marteau, Y. Déclais, D. Carbone, and E. Galichet, Geophysical muon imaging: Feasibility and limits, *Geophysical Journal International*, **183**(3), 1348–1361, (2010).
6. K. Nagamine, *Introductory Muon Science*, Cambridge University Press (2003).

7. H.K.M. Tanaka *et al.*, *Earth and Planetary Science Letters*, **263**, (2007).
8. K. Jourde, D. Gibert, and J. Marteau, Improvement of density models of geological structures by fusion of gravity data and cosmic muon radiographies, *Geoscientific Instrumentation, Methods and Data Systems*, **4**, 177–188, (2015a).
9. M. Rosas-Carbajal, K. Jourde, J. Marteau, S. Deroussi, J.-C. Komorowski, and D. Gibert, Three-dimensional density structure of la soufrière de Guadeloupe lava dome from simultaneous muon radiographies and gravity data, *Geophysical Research Letters*, **44**(13), 6743–6751, (2017).
10. K. Jourde, D. Gibert, J. Marteau, J. de Bremond d'Ars, and J.-C. Komorowski, Muon dynamic radiography of density changes induced by hydrothermal activity at the la soufrière of guadeloupe volcano, *Scientific Reports*, **6**, 33406, (2016b).
11. Y. Le Gonidec, M. Rosas-Carbajal, J. de Bremond d'Ars, B. Carlus, J.-C.Ianigro, B. Kergosien, J. Marteau, and D. Gibert, Abrupt changes of hydrothermal activity in a lava dome detected by combined seismic and muon monitoring, *Scientific Reports*, **9**(1), 1–9, (2019).
12. K. Jourde, D. Gibert, J. Marteau, J. de Bremond d'Ars, S. Gardien, C. Girerd, J.-C. Ianigro, and D. Carbone, Experimental detection of upward going cosmic particles and consequences for correction of density radiography of volcanoes, *Geophysical Research Letters*, **40**(24), 6334–6339, (2013).
13. H. Gómez, D. Gibert, C. Goy, K. Jourde, Y. Karyotakis, S. Katsanevas, J. Marteau, M. Rosas-Carbajal, and A. Tonazzo, Forward scattering effects on muon imaging, *Journal of Instrumentation*, **12**(12), P12018, (2017).
14. K. Jourde, D. Gibert, and J. Marteau, Joint inversion of muon tomography and gravimetry — a resolving kernel approach, *Geoscientific Instrumentation, Methods and Data Systems*, **5**, 83–116, (2015b).
15. H.K.M. Tanaka *et al.*, *American Journal of Science*, **308**, 843–850, (2008).
16. H.K.M. Tanaka, T. Uchida, M. Tanaka, M. Takeo, J. Oikawa, T. Ohminato, Y. Aoki, E. Koyama, and H. Tsuji, Detecting a mass change inside a volcano by cosmic-ray muon radiography (muography): First results from measurements at Asama volcano, *Japan, Geophysics Research Letter*, **36**, L17302, (2009).
17. N. Agafonova, *et al.* (OPERA Collaboration), *Observation of a first ν_τ candidate in the OPERA experiment in the CNGS beam*, arXiv:1006.1623, *Physics Letter*, **B691**, 138–145, (2010).

18. G. Boudon, A. Le Friant, J.-C. Komorowski, C. Deplus, and M. Semet, Volcano flank instability in the lesser antilles arc: diversity of scale, processes, and temporal recurrence, *Journal of Geophysical Research: Solid Earth*, **112**(B8), (2007).

19. M. Rosas-Carbajal, J.-C. Komorowski, F. Nicollin, and D. Gibert, Volcano electrical tomography unveils edifice collapse hazard linked to hydrothermal system structure and dynamics, *Scientific Reports*, **6**, 29899, (2016).

20. L. Bonechi, F. Ambrosino, L. Cimmino, R. D'Alessandro, G. Macedonio, B. Melon, N. Mori, P. Noli, G. Saracino, P. Strolin, F. Giudicepietro, M. Martini, M. Orazi, R. Peluso, The MURAVES project and other parallel activities on muon absorption radiography. *The European Physical Journal Conferences*, 02015, (2017) 10.1051/epj-conf/201818202015.

21. F. Beauducel *et al.*, The MU-RAY project: Summary of the round-table discussions, *Earth Planets and Space*, **52**, 145–151, (2010). doi:10.5047/eps.2009.07.003, 2010.

22. D. Carbone, D. Gibert, J. Marteau, M. Diament, L. Zuccarello, and E. Galichet, An experiment of muon imaging at Mt. Etna (Italy), *Geophysics Journal International*, **196**(2), 633–643, (2013).

23. D. Lo Presti, F. Riggi, and C. Ferlito, *et al.*, Muographic monitoring of the volcano-tectonic evolution of Mount Etna, *Scientific Report*, **10**, 11351, (2020). https://doi.org/10.1038/s41598-020-68435-y

24. V. Tioukov, A. Andrey, B. Cristiano, C. Lucia, N. D'Ambrosio, G. De Lellis, C. De Sio, F. Giudicepietro, G. Macedonio, S. Miyamoto, R. Nishiyama, M. Orazi, R. Peluso, A. Sheshukov, C. Sirignano, S.M. Stellacci, P. Strolin, and H.K.M. Tanaka, First muography of Stromboli volcano, *Scientific Reports*, **9**(6695), (2019).

25. H. Fujii, H. Kazuhiko, H. Kohei, K. Hidekazu, K. Hideyo, N. Kanetada, S. Kotaro, K. Shin-Hong, S. Atsuto, S. Takayuki, T. Kazuki, T. Fumihiko, T. Shuji, and Y. Satoru, Investigation of the Unit-1 nuclear reactor of Fukushima Daiichi by cosmic muon radiography, *Progress of Theoretical and Experimental Physics*, **2020**(4), 043C02, (2020). https://doi.org/10.1093/ptep/ptaa027

26. K. Jourde, J. Marteau, D. Gibert, J. Wassermann, and S. Gardien, J.-C. Ianigro, B. Carlus, C. Nussbaum, T. Theurillat, and C. Baumann, The Mont Terri udeground laboratory opalinus clay imaged by muon radiography and gravimetry, (2020), in preparation.

27. K. Jourde, D. Gibert, J. Marteau, J. de Bremond d'Ars, S. Gardien, C. Girerd, and J.-C. Ianigro, Monitoring temporal opacity fluctuations of large structures with muon radiography: A calibration experiment using a water tower, *Scientific Reports*, **6**(1), 1–11, (2016a).

28. M. Tramontini, M. Rosas-Carbajal, C. Nussbaum, D. Gibert, and J. Marteau, Middle-atmosphere dynamics observed with a portable muon detector, *Earth and Space Science*, **6**(10), 1865–1876, (2019). doi:10.1029/2019EA000655.

29. L.F. Thompson, J.P. Stowell, S.J. Fargher, C.A. Steer, K.L. Loughney, E.M. O'Sullivan, J.G. Gluyas, S.W. Blaney, and R.J. Pidcock, Muon tomography for railway tunnel imaging, *Physics Review Research*, **2**, 023017, (2020).

30. D.W. Schouten, and P. Ledru, Muon tomography applied to a dense uranium deposit at the McArthur River mine, *Journal of Geophysical Research: Solid Earth*, **123**, 8637–8652, (2018). https://doi.org/10.1029/2018JB015626

31. J. Marteau, D. Gibert, N. Lesparre, F. Nicollin, P. Noli, and F. Giacoppo, Muons tomography applied to geosciences and volcanology, *Nuclear Instruments and Methods in Physics A*, **695**, 23–28, (2012).

32. J. Marteau, J. de Bremond d'Ars, D. Gibert, K. Jourde, S. Gardien, C. Girerd, and J.-C. Ianigro, Implementation of sub-nanosecond time-to-digital convertor in field-programmable gate array: Applications to time-of-flight analysis in muon radiography, *Measurement Science and Technology*, **25**(3), 035101, (2014).

33. C. Cârloganu, V. Niess, S. Béné, E. Busato, P. Dupieux, F. Fehr, P. Gay, D. Miallier, B. Vulpescu, P. Boivin, C. Combaret, P. Labazuy, I. Laktineh, J.-F. Lénat, L. Mirabito, and A. Portal, Towards a muon radiography of the Puy de Dôme, *Geoscientific Instrumentation Methods and Data Systems*, **2**, 55–60, (2013). doi:10.5194/gi-2-55-2013.

34. N. Lesparre, J. Marteau, Y. Déclais, D. Gibert, B. Carlus, F. Nicollin, and B. Kergosien, Design and operation of a field telescope for cosmic ray geophysical tomography, *Geoscientific Instrumentation Methods and Data Systems*, **1**(1), 47–89, (2011).

35. S. Okubo, and H.K.M. Tanaka, *Measurement Science and Technology*, **23**, 042001, (2012).

36. A. Pla-Dalmau, A.D. Bross, and K.L. Mellott, Low-Cost Extruded Plastic Scintillator, *Nuclear Instruments and Methods in Physics Research A*, A, **466**, 482–491, (2001).

Chapter 8

Muography and Geology: Alpine Glaciers

Ryuichi Nishiyama

Earthquake Research Institute, The University of Tokyo
Albert Einstein Center for Fundamental Physics,
Laboratory for High Energy Physics, University of Bern
r-nishi@eri.u-tokyo.ac.jp

This chapter reviews the application of muography to alpine glaciers. The attenuation of muon flux observed beneath glaciers can be used to determine the shape of the boundary between the glacial ice and the underlying bedrock. The bedrock topography provides fundamental information on the erosion mechanism working beneath active glaciers, which could help to manage the risk of the global glacier recession.

1. Introduction

Muon absorption radiography (muography) provides information on the density distribution of the near-surface targets. The target of the survey ranges from volcanoes and natural caves (Chapter 7) to archeological sites (Chapter 9), etc. Muography is powerful when a strong density contrast is expected in the surveyed volume. In this respect, alpine glaciers are optimal targets, as the density varies greatly between ice ($\sim 0.9\,\mathrm{g\,cm^{-3}}$) and rock ($\sim 2.7\,\mathrm{g\,cm^{-3}}$).

Muography is capable of visualizing the shape of the boundary between the glacial ice and the underlying bedrock. The attenuation of the muon flux is known to depend on the column density traversed by muons. Here, the density-length X (called as column density or opacity in other literature) is defined as

$$X = \int \rho(\xi)d\xi, \tag{1}$$

where $\rho(\xi)$ is the density at the coordinate ξ and the integration is performed along the line of sight of the detector (Fig. 1). The flux of muons after passing through the density-length X of the target is calculated by integrating the energy spectrum of muons above the minimum energy $E_{\min}$ needed for muons to pass through X[1] hence,

$$F_\mu = \int_{E_{\min}(X)}^{\infty} \frac{dN(E, \theta)}{dE} dE, \qquad (2)$$

where dN/dE is the differential energy spectrum and the zenith angle θ represents the direction of incoming muons. When the surveyed volume is composed of ice and rock components, the density-length is written as

$$X = \rho_{\text{rock}} L_{\text{rock}} + \rho_{\text{ice}} L_{\text{ice}}, \qquad (3)$$

where L_{rock} and L_{ice} are the length of the rock and of the ice components, respectively. The sum of the two lengths $L(= L_{\text{rock}} + L_{\text{ice}})$ has to be known beforehand by means of a precise topography model. Since the density-length X is determined from the attenuation of the muon flux via Eq. (2), L_{rock} and L_{ice} are subsequently determined from Eq. (3) when ρ_{rock} and ρ_{ice} are given. A three-dimensional (3D) reconstruction of the boundary between ice and rock can be obtained from muon absorption along the different lines of sight of the detectors.

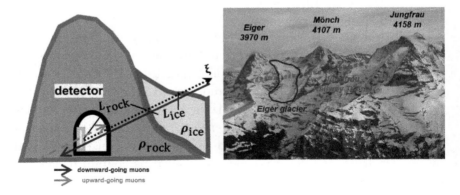

Figure 1. (Left) A schematic illustration of the muography of alpine glaciers. (Right) Photograph of Eiger glacier taken from the west.

Physicists and geologists of the University of Bern recognized the potentiality of the muography in geophysical exploration of alpine glaciers and they proposed an interdisciplinary novel project (Eiger-μ project, SNSF grant no. 159299). To reconstruct the bedrock profile of the Swiss Alpine glaciers, the group performed a measurement campaign in the Jungfrau region (Switzerland). The survey targets were the uppermost part of the Aletsch glacier[2] and the Eiger glacier.[3] In these studies, emulsion films (see Chapter 4) were employed as muon detectors. The emulsion films are special photo-graphic films that record the trajectories of charged particles passing through them. They are particularly suitable for measurements under the harsh environmental conditions typical of high mountain regions, although they require optical readout labor in the laboratory. This chapter reviews the technical issues related to the peculiar situation of glaciers and data analysis, through the examples of the measurements at the Eiger glacier.

2. Emulsion Detectors and Field Observations

Eiger glacier, the survey target of muography, is situated in the Central European Alps (Switzerland). The nearly $2.1\,\text{km}^2$ large glacier originates on the western flank of Mt. Eiger at about $3,600\,\text{m}$ a.s.l. From this height, it flows on the flank between Mts. Eiger and Mönch and terminates at about $2,400\,\text{m}$ a.s.l (Fig. 1). The surveyed region is the upper stream of the glacier ($>3,200\,\text{m}$ a.s.l.). Muon detectors were installed at three sites along the railway tunnel (Jungfrau Railway) situated inside the mountain so that the detectors surround the Eiger glacier from the inside. At each detector site, emulsion films were installed on the wall of the tunnel or the train station.

The position and orientation of the detectors are important for determining the direction of muons arriving into the films. Thus, the installation was performed in two steps. First, the detector frame (without films) was mounted on the wall. At this step, the facing direction of the frame was adjustable thanks to the ball joint on the rear of the frame; the tilt and azimuth angles were measured

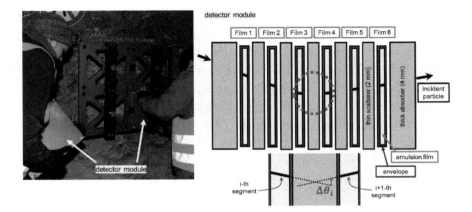

Figure 2. (Left) Detector frame installed vertically on the wall of the Jungfrau Railway. (Right) Schematic illustration of the detector module.

with a total station theodolite and adjusted to designated values (accuracy 0.2 degrees). After the orientation was fixed, emulsion films were transported from the laboratory and installed in the pockets on the frame (Fig. 2).

Emulsion films start recording trajectories of all the charged particles passing through them since their production. Consequently, special care is required to avoid combinatory backgrounds. Specifically, the films have been stored in an underground laboratory (30 m depth) and they were put in vacuum-seals just before use. During the transportation of the films from the laboratory to the experimental site, the films were shuffled to avoid accidental coincidence. Operators prepared the detector module on site just a few minutes before the installation. It should be also noted that temperature control is critical for emulsion films in general. If one leaves ordinary emulsion films at 30 degrees for more than 30 days, they become unanalyzable due to the signal fading and the increase of thermal noises.[4] However, this is not the case for glacier tunnels as the air temperature is very low, ~3 degrees throughout the year.

The physical background originates principally from the contamination of particles irrelevant to muography. Contamination of environmental radioactivity and contamination of low-energy cosmic rays are the most critical issues for emulsion films. The detector modules were designed to reduce these two types of backgrounds. The schematic illustration of the detector module is shown in Fig. 2.

Thick absorber plates (4-mm-thick stainless steel) were placed in front and behind the detector to absorb environmental radioactivity (alpha, beta, and gamma rays with sub-MeV to a few MeVs of energy). Alpha and beta rays are completely absorbed in these plates, while some gamma rays may penetrate the absorber and kick electrons near and inside the emulsion films. Nevertheless, these electrons would not be energetic enough to penetrate multiple films with straight trajectories.

Thin scatterer plates (2-mm-thick stainless steel) were inserted between each pair of adjacent films to deflect trajectories of low-momentum charged particles. The method of momentum measurement is described in the next section. Although lead or tungsten are more effective as scattering materials, stainless steel plates were employed for their cost and ease of handling.

The films remained in the tunnel for 3–5 months continuously. After extraction, they were developed in chemical baths and were scanned one by one with special automated optical microscopes. Details of the scanning microscopes at University of Bern can be found in the technical report.[5] The scanned data provided the position, direction, darkness of segmented tracks in each film. From segment data in individual films, a set of segments that penetrate multiple films with straight trajectories are reconstructed.[6] This reconstruction process includes the correction of alignment between adjacent films. The combinatory backgrounds are rejected in this process. Trajectories of physical backgrounds remain in a set of reconstructed tracks. However, they can be judged by the straightness of the tracks, defined in the next section, and rejected for further analysis.

3. Data and Analysis

This section focuses on the data analysis of the tracks reconstructed in the detectors. The three data sets, taken beneath the Eiger glacier, are introduced (detectors A, B, and C, see Fig. 3), to demonstrate the

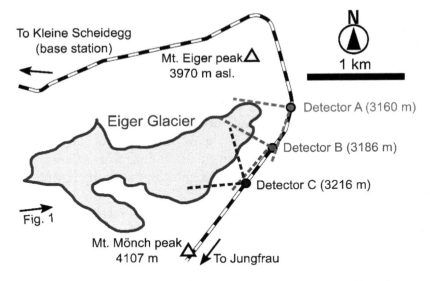

Figure 3.   Map of the surveyed target (Eiger glacier, Switzerland).

data quality and analysis methodology. For each site, eight modules
(25 cm × 20 cm × 6 layers) were installed. A total of ~7 m^2 films were
exposed.

Figure 4(a) represents the angular distribution of the reconstructed tracks in detector A. Since the films were installed vertically
on the wall of the tunnel, they record cosmic rays from forward
and backward directions simultaneously. The forward direction is
toward the Eiger glacier. The thickness of the mountain in the
backward direction is relatively thin, namely between 20 m and 40 m.
Background tracks can be originated from muons that entered the
mountain from the backward direction, scattered in the thin wall of
rock, and entered the detector with upward-going trajectories[7,8] (see
the illustration in Fig. 4). As emulsion films provide no time stamp on
incoming tracks, particles entering from the forward direction with
downward-going trajectories and those entering from the backward
direction with upward-going trajectories cannot be distinguished.

Track selection based on the particle momentum cut is then
needed to reject these background tracks. The idea is based on the
measurement of small deflection angles of trajectories traversing the

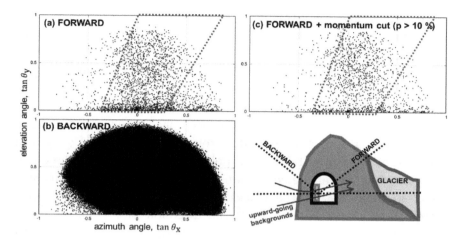

Figure 4. Angular distribution of observed tracks at detector A. Each dot in the plot corresponds to a single particle arriving at the detector (detector size: 1,512 cm^2, exposure duration: 107 days). (a) Tracks recognized as coming from the forward direction. (b) Tracks recognized as coming from the backward direction. (c) Same plot as (a) but with the momentum cut (see text). The blue dotted lines indicate the glacier in the field of view.

thin stainless plates. The scattering angle is described by a Gaussian distribution, with an RMS width given by

$$\theta_0 = \frac{13.6\,\text{MeV}}{\beta cp} z \sqrt{\frac{x}{X_0}} \left[1 + 0.038 \ln \left(\frac{xz^2}{X_0 \beta^2} \right) \right], \qquad (4)$$

where p, βc, and z are the momentum, velocity, and charge number of the incident particle, respectively, and x/X_0 is the thickness of the scattering medium in the unit of radiation length.[9] The RMS of the scattering angle of muons ($z = 1$) with $p = 0.5\,\text{GeV}/c$ after passing through 2 mm of steel is 8.6 mrad. As a typical angular resolution of the microscope is 5 mrad for tracks perpendicular to the film, this small scattering is barely observable. Quantitatively, an index of straightness of the track is defined as

$$\chi^2 = \sum_{i=1}^{n-1} \frac{(\Delta \theta_i)^2}{\sigma^2}. \qquad (5)$$

Here, n, $\Delta\theta_i$, and σ are the number of segments in the reconstructed track, the deflection angle from i-th to $i+1$-th segments (see Fig. 2), and the angular resolution, respectively. This index would follow a chi-square distribution when the momentum of incident particles is high enough. Thus, setting an appropriate χ^2 threshold, low-momentum tracks, having large scattering angles, are selectively removed from the analysis. Figure 4(c) shows the same angular distribution as Fig. 4(a), but after rejecting tracks with χ^2 values higher than the upper-tail critical point (10%). Comparing these two angular plots, it can be seen that the momentum cut significantly reduces the excess of the tracks near the horizon and does not affect those tracks coming from the direction of the glacier. The removed tracks mainly consist of low-energy (sub-GeV) particles and they are upward-going particles. The tracks near the horizon ($\tan\theta_y < 0.10 - 0.15$), therefore, were not used for the analysis of bedrock shape reconstruction.

Figure 5 shows the angular distribution of the reconstructed tracks in detector B. This site has a substantial thickness of rock in the backward directions (200–300 m) that is free of glacial ice. The flux data from these directions provided then the unique opportunity to calibrate the muon attenuation curve.

Figure 6 indicates the observed muon flux as a function of the thickness of the rock along the corresponding line of sight (closed squares). The curves are the theoretical prediction of flux attenuation assuming that the rock is occupying the full volume below the apparent surface, thus setting ice thickness $L_{\text{ice}} = 0$ in Eqs. (2) and (3) with the rock density value experimentally measured from rock samples, $\rho_{\text{rock}} = 2.68\,\text{g}\,\text{cm}^{-3}$. The muon energy spectrum model is taken from Tang et al. (2006).[10] The observed and predicted fluxes show an acceptable agreement. Also, one prominent characteristic that horizontal muons are more energetic is observed perfectly. A more quantitative test is done by taking the ratio of observed flux to the predicted one (inset Fig. 6). The ratio is found to be 0.94 ± 0.04, showing a slight deviation from identity but it remains constant as a function of the zenith angles. As a consequence of this test, a calibration factor $\alpha = 0.94$ was introduced and multiplied to

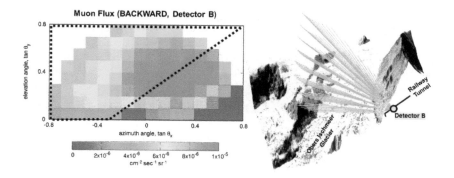

Figure 5. Flux of muons from the opposite side of the Eiger glacier, observed by detector B. The black dashed tetragon indicates the angular region that is free of glacial ice and is used for calibration analysis.

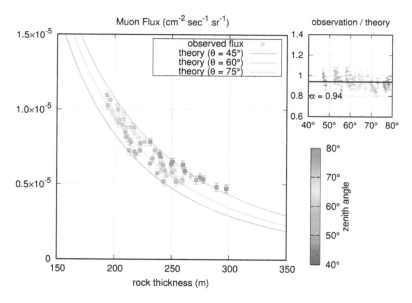

Figure 6. Calibration analysis. Muon flux of the pure rock region (dashed line in Fig. 5) as a function of rock thickness. The inset shows the ratio of the observed flux to the simulated flux.

the muon energy spectrum model in the analysis for bedrock shape reconstruction.

Figure 7(a) shows the flux of muons recorded by detector C. Detector C is closer to the glacier than the other two detectors.

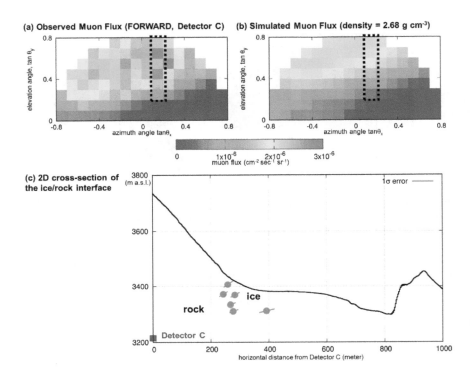

Figure 7. (a) Muon flux observed by detector C. (b) Simulated muon flux assuming the absence of the glacial ice. (c) 2D-cross-section showing the position of the ice/rock interface determined from the six angular bins enclosed by the black-dashed rectangle in (a) and (b).

From the detectors' perspective, the glacier lies almost entirely in the field of view. Figure 7(b) shows the theory prediction of the muon flux when assuming the absence of the glacier. The observed flux is higher than the simulated one by a factor of up to 2. This excess is due to the effects of the low-density glacial ice, allowing for higher transmittance of muons. The length of the ice component can be solved via Eq. (3), by assigning the density of glacial ice as $\rho_{ice} = 0.85\,\mathrm{g\,cm}^{-3}$.[11] For each bin in the angular plot, the position of the boundary is uniquely determined, as shown in the 2D cross-section (Fig. 7(c)). The uncertainty in the position of the boundary originates from statistical fluctuations on the number of muon events.

The position of the ice/rock boundary can be plotted in a 3D space for each rectangular bin. A 3D-visualization of the bedrock surface was obtained by combining all these point cloud data. Specifically, to render a surface from the point cloud, the points were multiplied to hundred synthetic datasets by adding a statistical fluctuation in each corresponding bin. The new points were then re-sampled in a cylindrical coordinate system with its axis parallel to the glacier flow direction. The choice of this coordinate system is due to the irregular shape of the Eiger glacier.

4. Results

Figures 8 and 9 show the tomographic images of the bedrock topography beneath the Eiger glacier. The resolution of the imaging ranges between 10 m and 30 m, depending on the coverage of the muon rays and the thickness of the overburden; when a thick rock is present in the field of view, the transmittance of muons decreases and corresponding statistical fluctuations of the number of muon events becomes severe. Since detector C is closest to the glacier, the bedrock in the corresponding region was well-constrained.

One relevant feature in Fig. 8 is the steep bedrock topography observed at the lateral margin of the glacier, as shown in the cross-sections (b) and (c). At the point α, the underlying bedrock is nearly vertical or maybe even overhanging (see also Fig. 7(c)), whereas in the other cross-sections ((a), (d), and (e)) the slope of the underlying bedrock gradually approaches the hillslope exposed above the glacier. The observed steep bedrock indicates that strong erosion occurs beneath active glaciers and such erosional effect could be highlighted. It should be noted that the steepest point α coincides with the crevasse (so-called bergschrund by alpinists) that separates the glacial ice and the rocky hillslope.

5. Discussion

The study performed at the Alpine glacier demonstrated for the first time that muography can be successfully used for 3D reconstruction of the boundary surface between glacier and rock. The accuracy

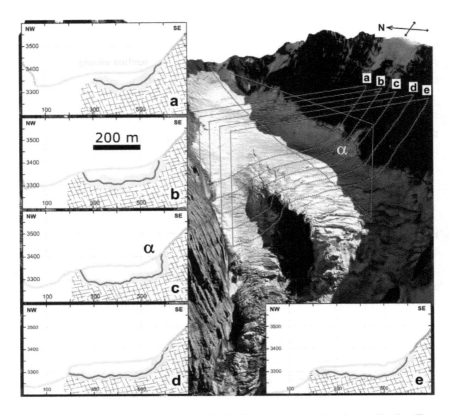

Figure 8. Tomographic images of the bedrock topography beneath the Eiger glacier (the glacier flows in the direction from (a) to (e). Blue curves represent the position of the best fit with surrounding cyan bands showing 1 sigma uncertainty.

of the reconstruction is assured by (i) background rejection, and (ii) calibration of muon absorption. The physical background noise has been a severe issue for volcano monitoring,[8,12,13] where the muon detectors are placed on the foot of the target mountains. In our case, the detectors were placed inside the mountain, thus the contamination of low-momentum particles was not severe for detectors B and C. Detector A was partly contaminated by upward-going particles[7] which enter the rear thin side of the mountain (20–40 m). The track selection based on momentum reduces a substantial amount of such tracks (but not all). The calibration analysis using

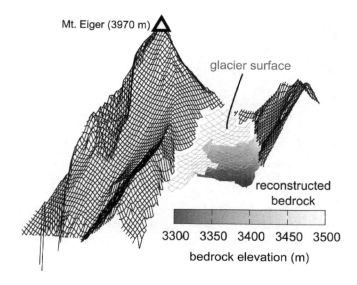

Figure 9. 3D representation of the reconstructed bedrock beneath the Eiger glacier. The vertical scale is exaggerated.

muons passing through the pure rock ensures the validity of the muon spectrum model employed in the analysis. The ratio of the observed flux to the one from theoretical prediction is found to be 0.94 ± 0.04 from the backward data of detector B. The slight deviation from the identity could be attributed to the systematic uncertainty due to the inefficiency of track reconstruction and the poor understanding of the muon energy spectrum model.

In the case of the Eiger glacier explained above, a nearly over-hanging bedrock topography of the underlying bedrock was revealed. Indeed, such a characteristic landscape is found in many places in Switzerland, where the glacier had set in the past and subsequently retreated (e.g., Fig. 10). As for the current worldwide recession of glaciers, a lot of attention is paid to predicting the potential collapse of the bedrock, where the vanishing ice decreases the stability of the bedrock. Muography could provide precious information on the underlying bedrock topography beneath remote and inaccessible glaciers. The information could be useful to estimate and manage the risk of infrastructures in Alpine regions.

Figure 10. Vertical and overhanging cliff exposed after the last deglaciation (First, Grindelwald, Switzerland).

6. Future Prospects

Now that the applicability of muography to bedrock reconstruction has been demonstrated with emulsion detectors, the next step would be monitoring of the dynamics in glacial processes. An immediate application could be the use of muography as remote snow gauges. The mass of the glacial ice varies over seasons; it increases by snowfall in winter and decreases by melting in summer. This mass balance of the glacial ice could be monitored with muography, for example by comparing the seasonal average of the muon flux observed underneath the glaciers.

A more challenging target could be the detection of water flow channels inside the glaciers.[14] Figure 11 illustrates how the water molten from the glacial ice flows inside the glacier. At the beginning of the melting season, the meltwater flows through tiny gaps or cavities between the ice and the underlying bedrock (Phase 1). As the volume and pressure of the water increase, the water begins to flow inside the body of the glacier via ice-walled conduits (Phase 2). In the end of summer, when the melting ceases, the channel becomes empty and remains empty for a few weeks before being closed by the creep of the ice (Phase 3). If muon detectors were placed underneath such

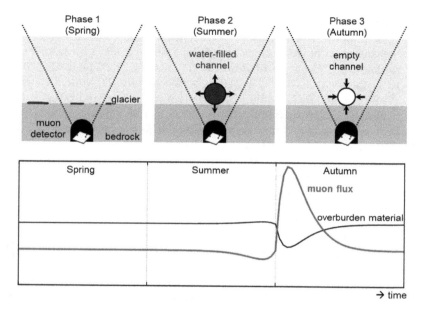

Figure 11. Schematic illustration of seasonal evolution of englacial channels (top) and the corresponding muon flux variation (bottom).

channels, the variation of the muon flux corresponding to each phase would be observed. The maximum of the muon flux would be seen in Phase 3, because the density-length decreases due to empty channels. The understanding of the morphology of those englacial conduits would benefit constraining the sliding velocity of glaciers,[15] the erosion power of glaciers,[16] and the occurrence of glacial outbursts.[17] Although emulsion films were the best choice for static measurements in cold and remote places, they do not provide the timing information on muons' arrival. This sort of challenging measurement would necessitate digital muon detectors (plastic scintillation detectors or gaseous detectors) which work under the extremely harsh and cold environments of alpine regions.

References

1. D.E. Groom, N.V. Mokhov, and S.I. Striganov, Muon stopping power and range tables 10 MeV–100 TeV, *Atomic Data and Nuclear Data Tables*, **78**, 183–356, (2001).

230   R. Nishiyama

2. R. Nishiyama, A. Ariga, T. Ariga, S. Käser, A. Lechmann, D. Mair, P. Scampoli, M. Vladymyrov, A. Ereditato, and F. Schlunegger, First measurement of ice-bedrock interface of alpine glaciers by cosmic muon radiography, *Geophysical Research Letters*, **44**(2), 6244–6251, (2017).
3. R. Nishiyama, A. Ariga, T. Ariga, A. Lechmann, D. Mair, C. Pistillo, P. Scampoli, P.G. Valla, M. Vladymyrov, A. Ereditato, and F. Schlunegger, Bedrock sculpting under an active alpine glacier revealed from cosmic-ray muon radiography, *Scientific Reports*, **9**, 6970, (2019).
4. A. Nishio, K. Morishima, K. Kuwabara, T. Yoshida, T. Funakubo, N. Kitagawa, M. Kuno, Y. Manabe, and M. Nakamura, Nuclear emulsion with excellent long-term stability developed for cosmic-ray imaging, *Nuclear Instruments and Methods in Physics Research A*, **966**, 163850, (2020).
5. A. Ariga, T. Ariga, A. Ereditato, S. Käser, A. Lechmann, D. Mair, R. Nishiyama, C. Pistillo, P. Scampoli, F. Schlunegger, and M. Vladymyrov, A nuclear emulsion detector for the muon radiography of a glacier structure, *Instruments*, **2**, 7, (2018).
6. V. Tioukov, I. Kreslo, Y. Petukhov, and G. Sirri, The FEDRA — Framework for emulsion data reconstruction and analysis in the OPERA experiment, *Nuclear Instruments and Methods in Physics Research A*, **559**, 103–105, (2006).
7. K. Jourde, D. Gibert, J. Marteau, J. de Bremond d'Ars, S. Gardien, C. Girerd, J.-C. Ianigro, and D. Carbone, Experimental detection of upward going cosmic particles and consequences for correction of density radiography of volcanoes, *Geophysical Research Letters*, **40**, 6334–6339, (2013).
8. R. Nishiyama, A. Taketa, S. Miyamoto, and K. Kasahara, Monte Carlo simulation for background study of geophysical inspection with cosmic-ray muons, *Geophysical Journal International*, **206**, 1039–1050, (2016).
9. P.A. Zyla, *et al.*, (Particle Data Group), The review of particle physics, *Progress of Theoretical and Experimental Physics*, 083C01, (2020).
10. A. Tang, G. Horton-Smith, V.A. Kudryavtsev, and A. Tonazzo, A Muon Simulations for Super-Kamiokande, KamLAND and CHOOZ, arXiv:hep-ph/0604078, (2006).
11. M. Huss, Density assumptions for converting geodetic glacier volume change to mass change, *The Cryosphere*, **7**, 877–887, (2013).
12. D. Carbone, D. Gibert, J. Marteau, M. Diament, L. Zuccarello, and E. Galichet, An experiment of muon radiography at Mt Etna (Italy), *Geophysical Journal International*, **196**(2), 633–643, (2014).
13. F. Ambrosino, A. Anastasio, A. Bross, S. Béné, P. Boivin, L. Bonechi, C. Cârloganu, R. Ciaranfi, L. Cimmino, Ch. Combaret, *et al.*, Joint measurement of the atmospheric muon flux through the Puy de

Dôme volcano with plastic scintillators and Resistive Plate Chambers detectors, *Journal of Geophysical Research: Solid Earth*, **120**, 7290–7307, (2015).

14. H. Röthlisberger, Water pressure in intra- and subglacial channels, *Journal of Glaciology*, **11**(62), 177–203, (1972).

15. C. Schoof, Ice-sheet acceleration driven by melt supply variability, *Nature*, **468**, 803–806, (2010).

16. F. Herman, F. Beaud, J.D. Champagnac, J.M. Lemieux, and P. Sternai, Glacial hydrology and erosion patterns: A mechanism for carving glacial valleys, *Earth and Planetary Science Letters*, **310**(3–4), 498–508, (2011).

17. W. Haeberli, A. Kääb, D.V. Mühll, and P. Teysseire, Prevention of outburst floods from periglacial lakes at Grubengletscher, Valais, Swiss Alps, *Journal of Glaciology*, **47**(156), 111–122, (2001).

https://doi.org/10.1142/9789811264917_0009

Chapter 9

Muography and Archaeology

Kunihiro Morishima

Nagoya University,
Furo, Chikusa, Nagoya, Aichi 464-8602, Japan
morishima@nagoya-u.jp

In 1967, Alvarez *et al.* applied cosmic ray muon radiography to explore the hidden chamber inside Khafre's pyramid in Egypt. This was the first attempt to take a two-dimensional (2D) transmission image based on the density contrast inside the target using cosmic rays; however, no new space was discovered inside the pyramid. Approximately 50 years later, ScanPyramids conducted a survey of Khufu's Pyramid and discovered a huge unknown space named ScanPyramids' Big Void inside the pyramid. This discovery proved that cosmic ray muon radiography is extremely effective in an investigation of archeological sites. In addition to the Egyptian pyramids, the Pyramid of the Sun of the ancient Mesoamerican civilization, the underground archeological sites in ancient Italy, and the Japanese mounds called kofun have also been explored. With its development, cosmic ray muon radiography technology together with further collaboration between archeologists in the humanities and sciences is expected to become a standard technology for the investigation and protection of archeological sites.

1. Introduction

Cosmic ray muon radiography is a non-destructive imaging technique that visualizes the density contrast inside objects up to several kilometers thick by using the high penetrating power of muons, which are generated when high-energy cosmic rays collide with the Earth's atmosphere. This technique has been used for various investigations, including archeological sites, volcanoes,[1] and nuclear reactors.[2, 3] Classical surveying methods of archeological sites have relied on historical documents and the hypotheses based on them, resulting in a direct destruction of the sites. With the development of science and technology, non-destructive inspection techniques such as electromagnetic radar surveys (e.g., ground-penetrating radar) and micro-gravimetry have been used as supplementary methods to narrow down the area to be investigated and improve the efficiency of an excavation; however, these techniques also have their own problems. In electromagnetic radar surveys, clear images cannot be obtained because the radar detects reflections from the boundary surfaces with different electrical properties, and the survey depth is limited to a short distance (e.g., within 1 m). Measurements using microgravimetry are insufficient in terms of resolution. Cosmic ray muon radiography, owing to its measurement principle, can visualize a wide angular range at a time as a high-resolution density contrast image, even for 100 m scale archeological monuments. Thus, the use of cosmic ray muon radiography, which is completely different from conventional techniques in terms of depth and quality, can contribute to important archeological discoveries.

In this chapter, an overview of the archeological surveys carried out is given using cosmic ray muon radiography, without regard to chronological order.

2. Egyptian Pyramids

The pyramids, built in the Nile River basin in Egypt, are the world's oldest and largest stone structures, using limestone as the main building material. The role of the pyramids and how they were built are still topics of significant interest, and archeological excavations and investigations using the latest non-destructive technologies have been conducted on pyramids of various sizes. In the history of pyramid surveys, cosmic ray muon radiography, which was conducted twice during investigations into the pyramids at Giza during a 50-year period, is a technologically innovative method, and the results obtained from recent investigations are extremely important from an archeological perspective.

2.1. *Khafre's Pyramid*

The first survey of an archeological site using cosmic ray muon radiography was conducted by Alvarez *et al.* on Khafre's Pyramid in Egypt.[4] This pyramid is one of the three great pyramids built at Giza near Cairo at the time of the 4th Dynasty during the Old Kingdom period in the 25th century BCE, along with the pyramids of Khafre's father, King Khufu, and son, King Menkaure. Alvarez focused on the differences in the internal structures of the pyramids of Kings Khufu and Khafre, which are the largest of the Egyptian pyramids (Fig. 1). Although there is only one chamber at the base of Khafre's Pyramid, the interior of Khufu's Pyramid has a complex internal structure with three chambers and one large corridor arranged three-dimensionally (3D) through passageways. Alvarez hypothesized that a chamber of the same size as that in Khufu's Pyramid might also exist in Khafre's Pyramid and devised a method to investigate the interior of the huge stone structure by using cosmic ray muons to visualize it as applied in X-ray radiography.

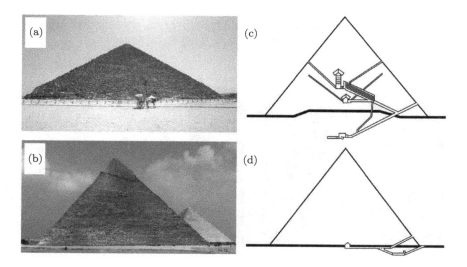

Figure 1. Pyramids at Giza: (a) Khufu's Pyramid, (b) Khafre's Pyramid, (c) cross-sectional view of Khufu's Pyramid, and (d) cross-sectional view of Khafre's Pyramid.

In 1967, Alvarez *et al.* set up a spark chamber in the Belzoni Chamber, located at the bottom of Khafre's Pyramid, and measured the directional distribution of muons that passed through the stones of the pyramid and reached the detector to search for an unknown space located above the detector. This experiment was not only the first application of muon radiography to an archeological site survey but also the first attempt of this kind. The detector consisted of two spark chambers of $1.8\,\text{m}^2$ per side and three scintillation counters. The spark chambers were activated by signals from the scintillation counters located on both sides of the spark chambers, and iron was placed between the two scintillators located at the bottom to suppress the effects of muon scattering. This large detector occupied most of the southeastern part of the Belzoni Chamber, as shown in Fig. 2(a), but when it was brought into the pyramid, it was split into several pieces and assembled internally. One factor that made this experiment possible at the time was the invention of a spark chamber with a digital readout, and several months of observations were carried out starting in 1968, yielding 0.65 million reliable muon tracks (Fig. 2(b)). The analysis was conducted for an angular range of a cone within a radius of 35° with respect to the vertical upward direction of the detector. Based on a comparison of the directional

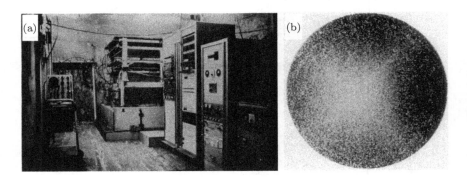

Figure 2. Observation of Khafre's Pyramid by Alvarez: (a) detector containing the spark chambers in the southeastern part of the Belzoni Chamber (modified version of Fig. 6 of Ref. 4) and (b) scatter plots of the observed data corrected for the geometrical acceptance of the detector (modified version of Fig. 13 of Ref. 4).

distribution of cosmic ray muons passing through the pyramid using simulations predicted from the known structure of the pyramid, it was confirmed that, in addition to the square shape of Khafre's Pyramid, the shape of the original limestone surface blocks at the top of the pyramid was also detected. However, within the observation area of the upper part of the Belzoni Chamber, which is equivalent to 19% of the pyramid volume, as investigated by Alvarez *et al.*, no space of the size found in Khufu's Pyramid was identified.

2.2. *ScanPyramids*

In 2015, the international collaborative research project titled ScanPyramids[5] was launched to investigate the Egyptian pyramids using non-destructive state-of-the-art survey techniques such as cosmic ray muon radiography and infrared imaging. This project covers four pyramids: Khufu's Pyramid, Khafre's Pyramid, the Red Pyramid, and the Bent Pyramid. As of 2021, the Bent Pyramid and Khufu's Pyramid have been surveyed using cosmic ray muon radiography.

2.2.1. *Bent Pyramid*

In 2015, the technology was demonstrated on the Bent Pyramid at Dahshur, built by King Sneferu, the father of King Khufu, using the known chamber inside (Fig. 3). The Bent Pyramid has one entrance on the north side and one on the west side, with two chambers inside leading from each entrance. The purpose of this experiment was to visualize the upper chamber by observing the upper part of the pyramid from a relatively lower chamber and to explore the unknown space. At the time, it was difficult to supply power to the inside of the Bent Pyramid, and thus the observation was conducted by Nagoya University using nuclear emulsion plates,[6] which do not require a power supply.[7] In the past, the analysis of nuclear emulsion plates was time-consuming and could not be conducted on a large scale; however, the recent development of an automatic scanning system[8] has made it possible to apply large-scale cosmic ray muon radiography inside the pyramid. Forty sets of nuclear emulsion plates

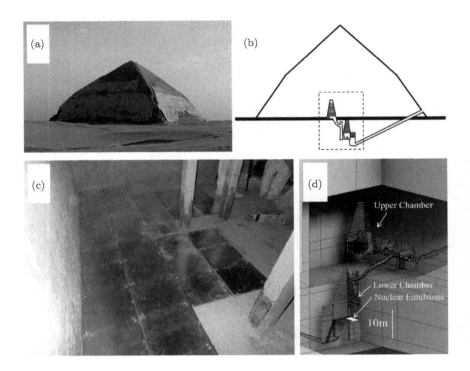

Figure 3. Observation of the Bent Pyramid at Dahshur (modified version of Fig. 3 from Ref. 7): (a) Bent Pyramid, (b) cross-sectional view of Bent Pyramid, where the red dots indicate the locations where the detectors were installed, (c) nuclear emulsion detectors in the lower chamber in Bent Pyramid, and (d) 3D enlargement of the area enclosed by the square dotted line in (b).

(25 cm × 30 cm) were installed in the lower chamber (total effective area of approximately 3 m^2), and observations were carried out for 40 days, and approximately 10 million tracks were detected. The upper chamber was visualized by comparing the angular distribution within ±45° from the detector with a simulation reflecting the 3D geometry of the pyramid based on a survey map (Fig. 4). However, no space of the same size as that of the upper chamber was detected.

2.2.2. *Khufu's Pyramid*

After the demonstration of the technology in the Bent Pyramid, an exploration of the unknown structure of Khufu's Pyramid was carried

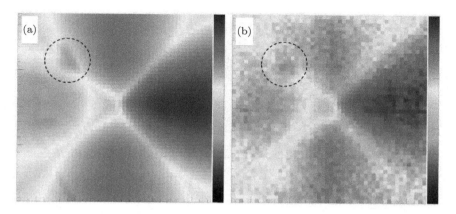

Figure 4. (a) 2D histograms of angular distribution of simulated muon flux and (b) observed muon flux. Red indicates a high muon flux, and blue indicates a low muon flux. The dotted circle corresponds to the upper chamber (modified version of Fig. 4 from Ref. 7).

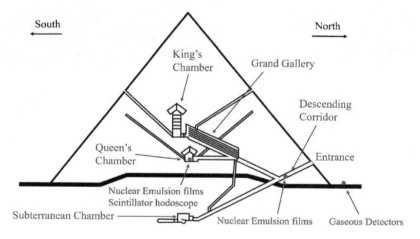

Figure 5. Cross-sectional view of Khufu's Pyramid. The internal structure and the location of the detector.

out using nuclear emulsion plates developed by Nagoya University, scintillation detectors developed by KEK, and gaseous detectors developed by CEA (Figs. 5 and 6).

In 2016, a team from Nagoya University observed cosmic ray muons (6.3 million muons) after installing three nuclear emulsion

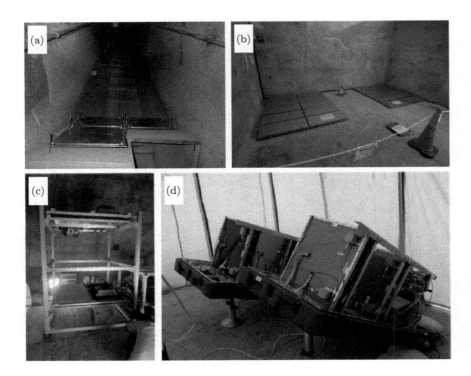

Figure 6. Detectors installed in Khufu's Pyramid (modified version of Fig. 1 from Ref. 9): (a) nuclear emulsion detectors installed in the Descending Corridor, (b) nuclear emulsion detectors installed in the Queen's Chamber, (c) a scintillation detector installed in the Queen's Chamber, and (d) gaseous detectors installed in front of the pyramid.

plates[6] (25 cm × 30 cm) for 67 days inside the Descending Corridor leading from the entrance to the south side of the pyramid of Khufu's Pyramid when it was first built (Fig. 6(a)).[7] Owing to their compactness, the nuclear emulsion plates were successfully installed within the narrow space of the Descending Corridor. As indicated by a comparison between the observations and the corresponding simulations, which was achieved by combining architectural drawings, photogrammetry, and laser scanner measurements for an angular range within ±45° from the detector, an excess region of muons extends along the north–south direction (Fig. 7). This indicates that there is a low-density region in this excess direction compared to the

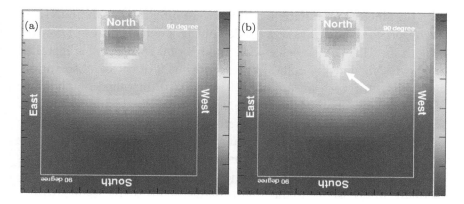

Figure 7. (a) 2D histograms of the angular distribution of the simulated muon flux and (b) the observed muon flux. Red indicates a high muon flux and blue indicates a low muon flux. A white arrow shows an excess muon region corresponding to the discovered corridor-like space, called SP-NFC (modified version of Fig. 6 from Ref. 7).

surrounding area, and this is the first study showing the existence of an unknown space in Khufu's Pyramid. Located just above the Descending Corridor, the space was thought to be a corridor-like structure continuing in the north–south direction and was therefore labeled the ScanPyramids' North Face Corridor (SP-NFC). This space is of significant archeological interest because it could be associated with a gable structure called Chevron on the north face of the pyramid.

In parallel, nuclear emulsion plates[6] were installed at two locations concurrently, one on the floor of the Queen's Chamber and one inside the narrow cave, called niche, approximately 1 m in both width and height located to the east of the Queen's Chamber.[9] Observations were conducted (Fig. 6(b)) from these locations, and it is expected that the two large spaces called the Grand Gallery and the King's Chamber, which are located above the Queen's Chamber, will be identified. Nuclear emulsion plates were periodically exchanged (retrieved and installed) inside the pyramid for continuous observation. The nuclear emulsion plates were developed in Egypt and were scanned at Nagoya University in Japan; in addition, the observation data obtained from different periods were integrated. An analysis of the observed cosmic rays was carried out in a detector area of $0.45\,m^2$ in both locations. A detector with an area of $0.75\,m \times 0.6\,m$ installed in a cave yielded 4.4 million tracks during 98 days of observation, and a detector with an area of $0.9\,m \times 0.5\,m$ installed on the floor of the Queen's Chamber yielded 6.2 million tracks during 140 days of observation, respectively, which were used for analysis. A comparison between the observation results (angular range within $\pm45°$ from the detector) and the results of a simulation corresponding to 1,000 days of observation showed the detection of an excess region with more muons than expected at both sites (Figs. 8(a)–8(c)). As a result of triangulation using the results of both locations, the position of the space responsible for the excess muons was confirmed to be just above the Grand Gallery.

Observations using scintillation hodoscopes[3] were conducted by moving the detector to two different locations, one on the east side and the other on the west side of the interior of the Queen's Chamber (Fig. 6(c)).[9] The scintillation detector had an area of 1.2 m $\times$ 1.2 m and the angular range was within $\pm34°$ to $\pm45°$ depending on the location. A multipixel photon counter sensor was used as photodetector. To pass through the narrow corridor into the chamber, the scintillation plates were designed as foldable structures and assembled inside the Queen's Chamber. Observations were made for 5 months (4.8 million events) on the east side of the Queen's Chamber and for 8 months (12.9 million events) on the west side. The observation results were compared with those of the simulations, and the results of the scintillation detector also confirmed the space discovered by the nuclear emulsion plates (Figs. 8(d)–8(f)). The results of the independent observations by these two different detector techniques were consistent, and the new space was found to be located just above the Grand Gallery (40 to 50 m above the Queen's Chamber) and more than 30 m long in the north–south direction. Furthermore, several muon observations revealed that the new space has a cross-sectional area similar to that of the Grand Gallery.

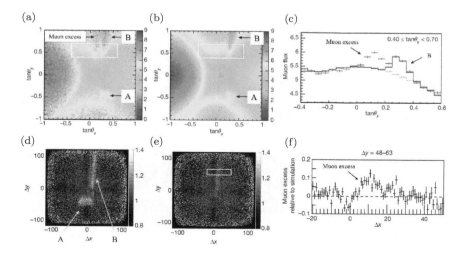

Figure 8. Results of observations conducted with nuclear emulsion detectors in the narrow corridor and scintillation hodoscopes on the west side in the Queen's Chamber (modified version of Figs. 2 and 3 from Ref. 9). Letter A in the figure indicates the King's Chamber and letter B indicates the Grand Gallery. (a) A 2D histogram of the angular distribution of the observed muon flux. (b) The simulated muon flux corresponding to (a). (c) A histogram of the angular region indicated by the white solid square line in (a) and (b). The vertical axis shows the muon flux. Data are shown in red. The simulation including the inner structures are shown by the solid black line. The simulation without the inner structures is shown by the grey dashed line. (d) A 2D histogram of angular distribution of the observed muon flux after normalization by the simulated muon flux without inner structures. (e) 2D histogram of angular distribution of the observed muon flux after normalization using the simulated muon flux with inner structures. (f) A histogram of the angular region indicated by the yellow solid square line. The vertical axis shows an excess of muons relative to those predicted by the simulation.

Two micro-pattern gaseous detectors (Micromegas) based on an argon mixture[10] were placed outside the pyramid by the CEA and were pointed in the direction of the discovered space for 2 months of observation (Fig. 6(d)).[9] The active area of the detectors was 50 cm × 50 cm with an angular range of ±45°. The observational data were transferred to the CEA in France through a 3G connection for analysis. A total of 12.9 million tracks were used for the analysis, and as a result, excess muons owing to the new space were detected along

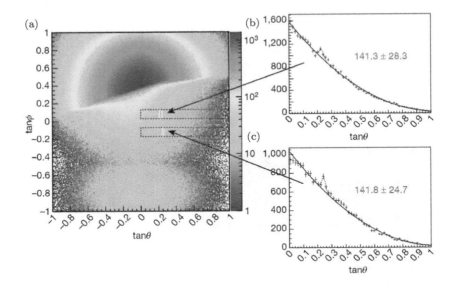

Figure 9. Results of observations conducted using gaseous detectors (modified version of Fig. 4 from Ref. 9). (a) A 2D histogram of the angular distribution of the observed muon flux at the logarithmic scale. (b), (c) Histograms of the angular region indicated by the dashed square line in (a). The regions with an excess of muons are indicated by the white solid square lines in (a). Excess muons correspond to the (b) new space and (c) Grand Gallery.

with a large corridor located at the center of the pyramid (Fig. 9). Thus, a huge new space named ScanPyramids' Big Void (SP-BV) was discovered inside Khufu's Pyramid through a data acquisition from three independent detectors and their analysis (Fig. 10).

The space discovered above the Descending Corridor and the space discovered above the Grand Gallery are archeologically remarkable, and the roles and relationships between these two spaces (SP-NFC and SP-BV) remain unclear. ScanPyramids has continued to make observations for clarifying the detailed 3D shapes and locations of these two spaces.

3. Ancient Mesoamerican Civilization

Among the ancient ruins of the Mesoamerican civilization, there are many huge pyramid-shaped stone structures like those of the

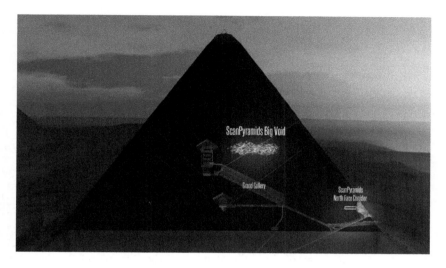

Figure 10. A schematic drawing of the cross section of Khufu's Pyramid and of the two spaces discovered inside it (credit: ScanPyramids). The space located at the center of the pyramid is the ScanPyramids' Big Void (SP-BV), and the space on the right side, corresponding to the north side, is the ScanPyramids' North Face Corridor (SP-NFC).

Egyptian pyramids. Like the Egyptian pyramids, they are well suited as targets for cosmic ray muon radiography, and observations have been made of the Pyramid of the Sun at the Teotihuacan in Mexico and a temple pyramid of the Mayan civilization at the Copan Ruinas in Honduras.

3.1. *Pyramid of the Sun*

Since 2011, cosmic ray muon radiography has been conducted by Aguilar *et al.* searching for hidden chambers inside the Pyramid of the Sun, which was built between the 1st and 2nd centuries CE at the Teotihuacan in Mexico.[11-14] Excavations of the Pyramid of the Sun have not revealed internal structures that are relatively common in prehistoric sites in Mesoamerica, such as those found near the Pyramid of the Moon. To observe cosmic rays, detectors were installed in a tunnel discovered 8 m beneath the Pyramid of the Sun, dug close to the symmetry axis (Fig. 11). The detectors

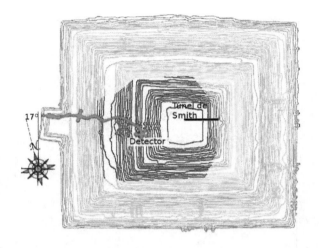

Figure 11. Contour map of the external shape of the Pyramid of the Sun. The tunnel where the detector is installed is shown in red, and the observation area is shown in dark blue (Fig. 1 of Ref. 14).

consisted of four scintillator plate detectors of 1 m × 1 m in size, allowing the muons to be identified and the background radiation removed, and six multi-wire proportional chambers to track the muons. The effective area of the detectors was 1 m², and the angular range was approximately ±50°. Observations were carried out over a period of approximately 2 years. The data were compared with the simulations incorporating the pyramid outline constructed through aerial surveying techniques, and the results were found to be in good agreement; however, no new chambers were identified.

3.2. Copan

Since 2018, a research group led by Nagoya University and Kanazawa University has been investigating the temple pyramids of the Copan civilization, which flourished between the 5th and 9th centuries CE, and is one of the most important sites of the Mayan civilization in Honduras. There was a culture of building stone chambers inside the temple pyramids of the Mayan ruins as burial spaces for kings, and several royal tombs have been discovered through tunnel excavations.

The Copan Ruins archeological site is one such example. During this project, the pyramid of Temple 11, which has not been sufficiently surveyed by tunnel excavations, has been the target of investigations, and cosmic ray muon radiography observations are being carried out using nuclear emulsion plates[6] to search for unknown royal tombs.

4. Ancient Civilization in Italy

In the underground areas of Italian cities, the remains of the ancient Greek and Roman eras between the 8th century BCE and the 6th century CE have been discovered, along with underground spaces created through the mining of stone for use in buildings. The technological development and investigation of cosmic ray muon radiography are underway in the subterranean remains of the ancient Roman era near Rome, Udine, and Naples, and at an ancient mine in Tuscany.

4.1. *Subterranean remains*

Menichelli *et al.* developed an underground buried-type detector for detecting underground cavities and structures by measuring the density distribution in the subsurface using cosmic ray muon radiography.[15, 16] They called their developed device MGR detector (Fig. 12), which has a cylindrical structure with two layers of scintillation fibers of 2 mm diameter and scintillation bars built into an aluminum cylinder of 2.24 m length and 14 cm diameter. Multi-anode photomultipliers were used as photodetectors. The detector can be inserted into a 20-cm diameter hole drilled into the ground at a maximum depth of 30 m. The performance of the detector was evaluated at two Italian archeological sites: Aquileia near Udine and the Traiano and Claudio ports near Rome. At the Aquileia site, the muon flux distribution (2 million events during a 70-h period), obtained by placing a detector 7 m underground, was compared with archeological maps and corresponded to roads and other structures. The results observed at the Traiano and Claudio ports were also compared with the geo-radar scan results and corresponded to high- and low-density regions.

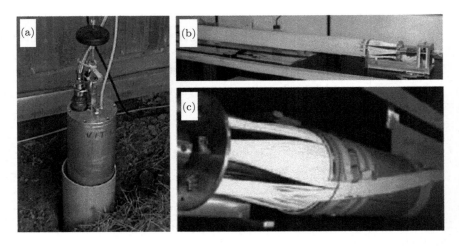

Figure 12.   MGR detector (modified version of Fig. 2 from Ref. 15 and Fig. 2 from Ref. 16). (a) Detector being inserted into the test hole. (b) Internal structures of the detector. (c) An enlarged view of the coil-shaped arrangement of the fibers.

4.2.   *Underground cavities*

Saracino *et al.* investigated the underground cavity of Mt. Echia, an ancient Neapolitan site dating from the 8th century BCE and detected unknown cavities estimating their 3D shapes.[17,18] Mt. Echia is composed of yellow tuff and has complex underground structures. To conduct cosmic ray muon radiography, detectors were installed at three locations in the underground cavity. MU-RAY detectors,[19] with an active area of 1.0 m × 1.0 m, and MIMA detectors,[20] with an active area of 0.4 m × 0.4 m, were used for this observation (Fig. 13). By comparing the results of the MU-RAY detector in the subsurface space with the drawings, it was confirmed that a known structure (a 4 m high space with a 40-m rock thickness) could be detected within a few hours of observation time. The density of the yellow tuff layer was estimated from the total observation data (approximately 14 million events were detected during 26 days of observation), and it was estimated to be 1.71 g/cm^3. In addition, a muon-excess region was detected in the observation results, which seemed to be an unknown structure. To investigate the unknown structure in more detail, a MU-RAY detector was newly installed at another location

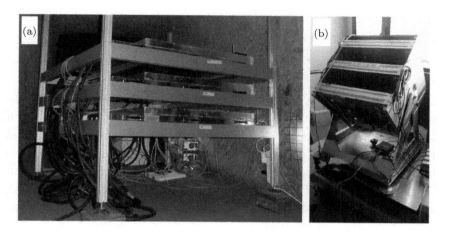

Figure 13. (a) MU-RAY detector and (b) MIMA detector (modified version of Fig. 2 from Ref. 18).

(where approximately 5 million events were detected during 8 days of observation), and a MIMA detector was installed at another location (where approximately 5 million events were detected during 50 days of observation). By analyzing these observation data together, a method was developed to estimate the location of the space in 3D.

4.3. *Ancient mines*

Measurements were conducted inside the Temperino mine, located in the San Silvestro archaeo-mining park in Tuscany.[21] There are tunnels where mining began during the Etruscan–Roman period. The aim of the measurements was to observe the space called the Gran Cava, an accessible excavation opening, and the detector was installed in a cave along the tunnel of the Temperino mine just below it to observe the upper area. The MIMA detectors[20] used for the observations were protected against moisture, and approximately 2 million events were detected for a period of 53 days. By analyzing these data and comparing them with simulations that consider the topography, which was achieved using a laser scanner (both the inside of the mine and the overlying hill), a map of the average density distribution within the observation area was obtained. The results suggest the presence of the Gran Cava, as well as unknown cavities and high-density rock veins.

5. Japanese Burial Mounds

Japanese ancient burial mounds of various sizes are located throughout the country and were built during the Tumulus period between the 3rd and 6th centuries CE. Particularly for the huge burial mounds, entry and excavation are prohibited or restricted. If the location and size of the stone chambers in the burial mounds can be determined, the style of the chambers will present a closer look at the time of construction and the specific image of the buried individual. For this purpose, cosmic ray muon radiography is considered to be effective, and several research groups in Japan have been investigating this approach. To date, groups from the Archaeological Institute of Kashihara, Nara Prefecture, and Nagoya University have confirmed the existence of a known cavity in the excavated Ishigami Tumulus using nuclear emulsion plates.[6, 22] Measurements are also underway at the unexcavated burial mounds of Kasuga Tumulus, Nishi-Norikura Tumulus, and Hashihaka Tumulus, some of which suggest the existence of stone chambers. The University of Tokyo, Kansai University, and others have been observing Imashirozuka Tumulus, Tsugeyama Tumulus, and Tsukuriyama Tumulus using gas detectors. A group from Okayama University and some other groups are planning to observe Tsukuriyama Tumulus and Ryoguzan Tumulus using scintillation detectors.

6. Summary

Cosmic ray muon radiography, which started with observations of Khafre's Pyramid by Alvarez and colleagues, has shown potential for archeological research with remarkable discoveries made at Khufu's Pyramid. This is a good example of how modern particle physics techniques can be used to bring completely new insights into archeological heritage. In addition to the Egyptian pyramids, the pyramids of the ancient Mesoamerican civilization, the underground ruins of the Roman era, and the Japanese burial mounds mentioned in this chapter have been applied to archeological sites all over the world, and several other research projects are currently underway. In the future, it will be important to select the best detector layout and an appropriate detector enabling the application of cosmic ray muon radiography at various archeological sites, depending on the shape and location of the target. In addition, by clarifying the detailed 3D structure of the discovered space, it is possible to conduct an archeological interpretation without an excavation and to minimize the damage and risk to a site during excavation. Furthermore, by improving the accuracy of the density measurement, it will be possible to evaluate the earthquake resistance from the perspective of cultural property protection.

In the future, the further technical development of cosmic ray muon radiography, along with the accumulation of archeological achievements through investigations into archeological sites fusing the humanities with the sciences, will allow such technology to become a standard in investigations into archeological sites and cultural properties, and open up new areas for its wider use.

References

1. H.K.M. Tanaka *et al.*, Development of an emulsion imaging system for cosmic-ray muon radiography to explore the internal structure of a volcano, Mt. Asama, *Nuclear Instruments and Methods A*, **575**, 489–497, (2007).
2. K. Morishima *et al.*, First demonstration of cosmic ray muon radiography of reactor cores with nuclear emulsion based on an automated high-speed scanning technology, *Proceedings of 26th Workshop on 'Radiation Detectors and Their Uses'* 27–36 (2012).
3. H. Fujii *et al.*, Detection of on-surface objects with an underground radiography detector system using cosmic-ray muons, *Progress of Theoretical and Experimental Physics*, **5**, 053C01, (2017).
4. L.W. Alvarez *et al.*, Search for hidden chambers in the pyramids, *Science*, **167**, 832–839, (1970).
5. http://www.scanpyramids.org/
6. K. Morishima *et al.*, Development of nuclear emulsion for muography, *Annals of Geophysics*, **60**, 0112, (2017).
7. K. Morishima *et al.*, Observation of cosmic rays with nuclear emulsions inside Egyptian pyramids, *Proceedings of the 35th ICRC*, **295**, (2017).
8. M. Yoshimoto *et al.*, Hyper-track selector nuclear emulsion readout system aimed at scanning an area of one thousand square

meters, *Progress of Theoretical and Experimental Physics*, **10**, 103H01, (2017).

9. K. Morishima *et al.*, Discovery of a big void in Khufu's Pyramid by observation of cosmic-ray muons, *Nature*, **552**, 386–390, (2017).

10. S. Bouteille *et al.*, A Micromegas-based telescope for muon tomography: The WatTo experiment, *Nuclear Instruments and Methods in Physics Research A*, **834**, 223–228, (2016).

11. R. Alfaro *et al.*, Searching for possible hidden chambers in the Pyramid of the Sun, *Proceedings of the 30th ICRC*, **5**, 1265–1268 (2007).

12. S. Aguilar *et al.*, Searching for cavities in the Teotihuacan Pyramid of the Sun using cosmic muons, *Proceedings of the 32th ICRC*, **4**, 317, (2011).

13. S. Aguilar *et al.*, Search for cavities in the Teotihuacan Pyramid of the Sun using cosmic muons: Preliminary results, *Proceedings of the 33th ICRC*, **364**, (2013).

14. S. Aguilar *et al.*, Search for cavities in the Teotihuacan Pyramid of the Sun using cosmic muons: Preliminary results, in *X Latin American Symposium on Nuclear Physics and Applications (X LASNPA)*, Montevideo, Uruguay (December, 2013).

15. M. Basseta *et al.*, MGR: An innovative, low-cost and compact cosmic-ray detector, *Nuclear Instruments and Methods in Physics Research A*, **567**, 298–301, (2006).

16. M. Menichellia *et al.*, A scintillating fibres tracker detector for archaeological applications, *Nuclear Instruments and Methods in Physics Research A*, **572,** 262–265, (2007).

17. G. Saracino *et al.*, Imaging of underground cavities with cosmic-ray muons from observations at Mt. Echia (Naples), *Scientific Reports*, **7**, 1181, (2017).

18. L. Cimmino *et al.*, 3D Muography for the search of hidden cavities, *Scientific Reports*, **9**, 2974, (2019).

19. F. Ambrosino *et al.*, The MU-RAY detector for muon radiography of volcanoes, *Nuclear Instruments and Methods in Physics Research A*, **732**, 423–426, (2013).

20. G. Baccani *et al.*, The MIMA project. Design, construction and performances of a compact hodoscope for muon radiography applications in the context of archaeology and geophysical prospections, *Journal of Instrumentation*, **13**, P11001, (2018).

21. G. Baccani *et al.*, Muon radiography of ancient mines: The San Silvestro Archaeo-Mining Park (Campiglia Marittima, Tuscany), *Universe*, **5**(1), 34, (2019).

22. K. Ishiguro, Study of Japanese Tumulus by muon-radiography, *Bulletin of The Society of Scientific Photography of Japan*, **81**(3), 258–262 (2018).

© 2023 World Scientific Publishing Company
https://doi.org/10.1142/9789811264917_0010

Chapter 10

Civil and Industrial Applications of Muography

D. Pagano* and A. Lorenzon[†]

*Department of Mechanical and Industrial Engineering,
University of Brescia, Italy
[†]Department of Physics and Astronomy,
University of Padova, Italy

Starting from the mid-1950s, techniques based on the absorption or scattering of cosmic ray (CR) muons, collectively referred to as muography, have been applied beyond the research domain only. Since the first pioneering measurement of the overburden of a railway tunnel, performed by E.P. George in 1955, applications of muography have steadily grown in number, especially in recent years. In this chapter, an overview of the most important civil and industrial applications of muography is given. Section 2 deals with applications of CR muons for the subsurface exploration, relevant in the mining industry, and for the monitoring of tunnels and other underground civil structures. Section 3 presents the use of muography for the inspection of large and/or inaccessible structures, such as a blast furnace and a nuclear reactor. Section 4 is dedicated to applications to nuclear controls, such as nuclear waste control, the inspection of dry storage casks for spent nuclear fuel, and vehicle and cargo scanning for nuclear contraband. Finally, Section 5 deals with the use of CR muons for the stability monitoring of civil and industrial structures, a technique generally referred to as muon metrology.

1. Introduction

Although the discovery of cosmic rays (CRs) dates back to the beginning of the 20th century, it was only from the mid-1950s that their use went beyond the research domain. In 1955, by measuring the CR muon intensity attenuation, E.P. George estimated the overburden of a railway tunnel,[1] kicking off what is generally referred to as *muon geotomography* (see Section 2) nowadays. Since then, techniques based on the absorption or scattering of CR muons have been successfully used in a plethora of scenarios: study of volcanoes, nuclear waste control, search for underground cavities, mineral exploration, monitoring of carbon dioxide geo-storage, bedrock profiles in glaciers, and much more.

In this chapter, an overview of the most important civil and industrial applications of muography is given: Section 2 presents industry-relevant applications of muon geotomography, which was already introduced in Chapter 6; Section 3 is dedicated to the use of muography for the inspection of large and inaccessible civil and industrial structures, and for exploration of the inner structure of buildings and industrial equipment; Section 4 provides an overview of the applications related to nuclear controls, such as the inspection of dry storage casks and vehicle and cargo scanning for nuclear contraband; finally, Section 5 presents the use of CR muons for the stability monitoring of structures, a technique usually referred to as *muon metrology* nowadays.[2–4]

2. Muon Geotomography for Subsurface Exploration

When designing new civil engineering structures, the knowledge of the subsurface composition at the construction site is crucial as it drives the choice of foundation type and depth, and determines their load-bearing capacity.[5] Subsurface exploration is also key for the mining industry, which is constantly looking for new deposits to keep pace with the increasing demand for metals and minerals.[6]

Common techniques used for the subsurface exploration include the use of electromagnetic, gravimetric, and magnetic sensors, as well as direct soil sampling.[7] In some cases, though, these methods end up being inadequate for the task because of the small coverage, low resolution, or high costs.[8] In recent years, imaging techniques of the subsurface density, based on the absorption of CR muons and collectively referred to as muon geotomography, have been investigated.

2.1. *Mining industry*

In 2010, a proof-of-principle trial was conducted at the Nyrstar Myra Falls mine in Canada.[9] The authors investigated the Price formation, which is a zinc-rich volcanic-hosted massive sulphide (VMHS) deposit, for the following reasons: (i) it was close to the surface (depth of $\sim$70 m), (ii) drill core data of the deposit were available, and (iii) tunnels for the placement of the muon detectors were already present. For the study, a single muon tracking detector was placed, in turn, at seven different locations, collecting data for about 15–20 days per site. Anomalies in density distribution, due to the Price deposit, were reconstructed in approximately the same locations by both drill core and muon data, with differences in the center of masses of 20 m at most. Unsurprisingly, reconstructed anomalies from muon data were found to be spatially larger than what was expected from the simulation based on drill core data: only ore of economic value was accounted in the simulation, whereas the deposit was actually surrounded by ore enriched rock, resulting in a larger footprint.[10]

In 2015, Bryman *et al.*[11] presented the results from the first blind test of muon geotomography applied to mineral exploration. The study was conducted at the Pend Oreille mine in USA, using the MX700 tunnel at 490 m below the surface. No prior information about the existence of ore bodies in the region under investigation was provided before the muon geotomography survey was completed. A large region with an excess of high-density material, with respect to the assumption of a uniform host rock density of 2.79 g/cm^3, was identified and found to be consistent with the ore model derived from drill core data. Data taking was performed at two different locations and lasted up to 5 months. As for the test at the Nyrstar Myra Falls mine, the reconstructed high-density region was larger than what was predicted by the drill core model ($\sim$20% lager in this case), likely due to additional mineralization of surrounding rock, not included in the model. The survey at Pend Oreille represents the first example of the search for a possible ore body by means of muon geotomography, demonstrating the applicability of the technique to mineral exploration.

2.2. *Tunnels and other underground civil engineering structures*

Another novel application of muon geotomography to civil engineering consists of monitoring of the overburden rock during tunnel digging operations.[8] Indeed, despite the meticulous planning, drilling operations are subject to real-time changes in presence of unexpected geological features, such as local high-density regions, cavities, and unstable terrain. For this reason, some authors are investigating the possibility of monitoring the rock surrounding a tunnel boring machine (TBM), by directly mounting a muon detector on it.[12–14] Since 2018, a muon telescope has been installed on the TBM used in the excavation of Paris subway line 15, and it has been oriented so that its field of view (FOV) is in the direction of drilling. Because of the TBM motion, collected data consists of a sequence of 2D views of reconstructed densities, based on short periods of data taking, which are then combined to create a 3D density model of the ground surrounding the excavated tunnel. The authors aim at developing a monitoring tool for the early detection of anomalies in front of the TBM, to modify the drilling operations accordingly. Results from both simulation studies and experimental data seem promising, though the study is still ongoing at the time of writing.

Thompson *et al.*[15] used a portable CR muon tracking system for the muography of the whole overburden of a UK railway tunnel: the Alfreton Old Tunnel, which is a disused 770 m long tunnel in Nottinghamshire. The tracking system, housed in a commercially available van, consisted of two horizontal layers of EJ-200 plastic scintillator. The upper layer was segmented into six $90 \times 15 \times 4\,\mathrm{cm}^3$ rectangular bars, whereas the lower one was segmented into three $30 \times 30 \times 4\,\mathrm{cm}^3$ square paddles. The top layer was positioned 76 cm above the bottom one, resulting in a 100° FOV along the direction of the tunnel, and a 76° FOV along its width. The data-collection campaign lasted for 12 days, with shifts of up to 8 h per day. Measurements were taken along the entire tunnel at 10 m intervals (5 m in case of overburden regions of interest) and, at each position, 20 or 30 min of muon data were collected, together with pressure and temperature readings.

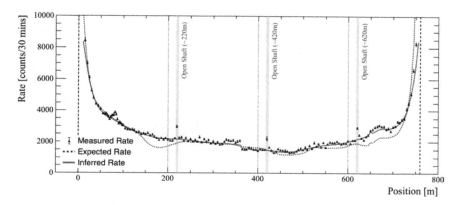

Figure 1. Measured and expected (from topographical information) muon flux rate as a function of the position in the tunnel.[15]

Figure 1 shows the variation of the measured muon flux rate as a function of the distance along the tunnel: in general, the observed rate is in good agreement with the expected one, derived from a topographical survey. Superimposed on the figure is also shown the location of the three known open shafts, represented as light gray boxes, which explains the corresponding sudden variation of measured rates. The authors investigated further some sectors of the tunnel, where significative (and unexpected) discrepancies between measured and expected rates were observed. In particular, the peak in data at around 80 m indicated an unknown hidden void, which was later confirmed by rail authorities.

2.3. *Portable borehole muon detectors for industrial and civil applications*

With the growing number of applications of muon geotomography, some research groups have started designing and developing compact borehole muon detectors (BMDs) for industrial and civil uses.

Oláh *et al.*[16] designed a BMD based on multi-wire proportional chambers: four vertically palced $20 \times 32 \, \text{cm}^2$ *close cathode chambers*[17] were horizontally stacked to create a 18 cm thick tracking system, providing an angular resolution of ~ 10 *m*rad. As variations of environmental parameters (such as temperature and pressure) can

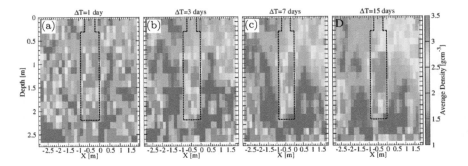

Figure 2. Reconstructed average soil density, as a function of the horizontal distance and depth, for a data-taking lasting 1 day (a), 3 days (b), 7 days (c), and 15 days (d). Dashed line indicates the position of the pillar. Figure courtesy of L. Oláh.

impact the gas quality and, consequently, the detection efficiency, a real-time compensation of the high-voltage was implemented. The detector was successfully used for the muography of an underground pillar: it was placed in a 3 meters deep well, at a distance of 3 m from an iron pillar with a concrete basement, buried for 2.2 m. Figure 2 shows the reconstructed average soil density, as a function of the horizontal distance and depth, for different data-taking periods. With 15 days of data-taking, the bottom of the pillar, at a depth of 2.2 m, was successfully observed and a spacial resolution of ~10 cm was reached.

For their 15 cm wide, 68 cm long, and 8 cm high BMD, Bonneville *et al.*[18] proposed instead a design based on scintillating rods with fiber readouts: four layers of 1 cm square polystyrene rods, coated with TiO_2 reflectors. A 2 mm wavelength shifting fiber was glued to each rod and read with a SiPM. The prototype was successfully tested in four different scenarios: (i) above ground, at both the University of Hawaii and Pacific Northwest National Laboratory (PNNL), (ii) in the Shallow Underground Laboratory at PNNL, at a depth of approximately 35 m-water-equivalent, and (iii) at high altitude (~2,300 m) at Los Alamos National Laboratory (LANL). Reconstructed images and measured fluxes, at all locations, were found to be in good agreement with dedicated Monte Carlo (MC) simulations. Moreover, at LANL, data were also acquired with a much larger muon detector: the Mini Muon Tracker (MMT),[19] consisting of 12 layers of drift tubes and covering an area of about 1.2 m^2 square. Despite the much higher resolution of the MMT, due to the larger detection area, flux plots from both the prototype and the MMT were found to be in very good agreement.

3. Muography of Civil and Industrial Structures

In addition to the study of natural structures, muography has also been applied for the inspection of large and/or inaccessible civil and industrial structures, as detailed in the next sections.

3.1. *Monitoring of blast furnaces*

In 2005, Nagamine *et al.* investigated the possibility of using muon radiography by absorption to inspect a blast furnace (BF) for iron making.[20] The experimental setup consisted of two side-to-side detectors, each made of two x–y sensitive layers of 10 cm × 1 m × 3 cm scintillating bars, for a total area of 2×1 m^2. Data were collected in two different configurations: (i) 20 days of data taking with the detection system placed at a height of 2.7 m from the ground and with a horizontal displacement of 17.3 m from the center of the furnace, and (ii) 25 days of data-taking with the detection system at a height of 4.7 m and the same horizontal distance from the furnace. With a total of 45 days of data-taking, the thickness of the base and side-walls of the furnace were measured with a precision of 5 cm, whereas for the local density of its iron-rich region, a precision of 0.2 g/cm^3 was reached. These results, although produced with long data-taking efforts, opened up to the possibility of a continuous monitoring of a furnace during its life cycle, something which has been investigated by other authors, too.

Åström *et al.* explored the capability of muon scattering tomography (MST) for imaging the different components (coke, burden and reduced metal) inside a blast furnace.[21] The study, based on MC simulations, considered two position and momentum sensitive detectors of large areas (5 m × 4 m each), to reduce the data-taking time to the order of few hours, which were placed at opposite sides of the furnace. Results showed that useful images to assess the structure of the components inside the BF could be obtained with a data-taking of ~8 hours. As the interior of a BF is a dynamic system, due to the burden downward movement, a data-taking lasting few hours can create movement blur and artifacts in reconstructed images. However, the authors showed they could mitigate the effect by transforming the input data as if it were collected by detectors synchronously moving with the burden.[4]

In another work,[22] the same authors used a detection system consisting of two 300 × 250 cm² drift chambers, made of 12 layers of drift tubes each, to reconstruct the linear scattering density (LSD), defined as the inverse of the radiation length, on a set of calibration samples (ferrous pellets and coke) and on experimental samples from the LKAB's blast furnace (located in Luleå, Sweden). Results showed not only that the muon scattering tomography was able to discriminate the ferrous pellets from the coke, but also that the LSD linearly correlated with the bulk densities of the measured materials, as shown in Fig. 3.

Gilboy *et al.*[23] proposed a completely different use of CR muons to assess the carbon liner thickness in a blast furnace, which is used for the thermal insulation of the steel vessel. The idea is to measure

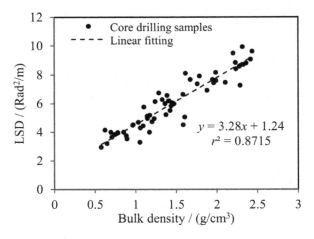

Figure 3. Correlation between LSD, as measured by means of muon scattering tomography, and the bulk densities of the core-drilling samples.[22]

the rate of neutrons, induced by absorption of CR negative muons, to distinguish between high-Z and low-Z materials, as the nuclear absorption of negative muons is approximately proportional to Z^4.[24] The potential of this approach has been investigated with a simple model of the furnace liner assembly: up to 8 carbon blocks were stacked in pairs on a wooden pallet and covered by 5 cm thick lead radiation shielding blocks, in either a single or a double layer. The thermal neutron intensity, detected by a He-3 tube, placed below the pallet, as a function of carbon liner thickness was measured and compared to dedicated MC simulations. The sensitivity of counting rate on the carbon thickness, as well as the general agreement between data and MC predictions, seem to support the proposed approach.

3.2. *Inspection of the Unit-1 reactor at Fukushima Daiichi Nuclear Power Plant*

On March 11, 2011, following the tsunami generated by a magnitude 9.0 earthquake, a nuclear crisis occurred at Fukushima Daiichi Nuclear Power Plant.[25] In December 2011, the Japanese government announced a cold shutdown of the plant and the beginning of a phase of cleanup and decommissioning. However, the dismantling of the reactors requires a realistic estimate of the extent of the damage to the cores, and, for example, for the Three Mile Island accident, it took more than 3 years before a camera could be put into the reactor. Several groups in both the United States and Japan realized, though, that muography could provide valuable information about the damaged cores.

In 2012, Borozdin *et al.*[26] compared the performance of muon scattering and transmission techniques in imaging the reactor, by means of MC simulations. A model of a boiling water reactor, similar to the Fukushima Daiichi Unit-1 reactor, was simulated in three different scenarios: (i) intact core, (ii) a core with a 1 m diameter void, and (iii) empty core. Plots of both scattering and transmission images are shown in Fig. 4 for exposure times from 1 h to 6 weeks. The results were produced for a 50 m^2 detector.

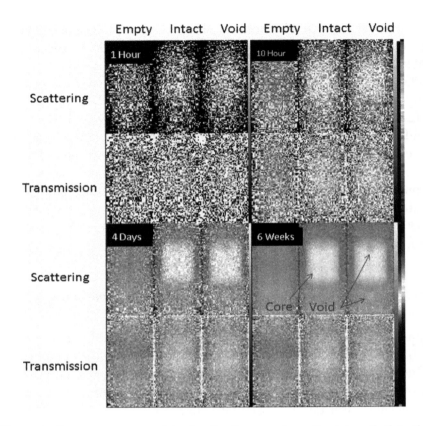

Figure 4. Reactor reconstructions for the three configurations described in the text, using scattering and transmission tomography.[26]

Differences in performance between the two techniques are evident: while the 1 m diameter void is already clearly visible after 4 days of data-taking with the muon scattering tomography, for the transmission technique even 6 weeks of data do not allow a reliable reconstruction of the void.

The team at Los Alamos National Laboratory also imaged a reactor mock-up with their Muon Mini Tracker (see Section 2.3).[27] The mock-up consisted of two layers of concrete blocks (with a thickness of 2.74 m each) and a lead assembly in between, which was arranged into different configurations. One detector was placed at a height of 2.5 m, whereas the other one was installed on the ground at the other side of the mock-up. In one of the configurations, the lead formed a conical void, similar in shape to the melted core of the Three Mile Island reactor. The authors showed they were able to reconstruct the void with a data-taking of 3 weeks.

A team of people from KEK, University of Tsukuba, and University of Tokyo, tested the potential of muon radiography to explore the inner structure of Fukushima Daiichi reactors, by imaging a GM MK-II-type nuclear reactor at the Japan Atomic Power Company (JAPC).[28] The detection system, consisting of eight 1×1 m^2 tracking planes, made of $9.7 \times 6 \times 1,000$ mm^3 scintillating bars, was installed in March 2012 at $\sim$64 m from the reactor center. During the so-called PERIOD-1 (up to June 18), the detector pointed toward the center of the fuel loading zone, whereas, during PERIOD-2 (from June 28 to October 5), it pointed toward the fuel storage pool. The authors showed they could reconstruct an image of the main structures of the reactor: containment vessel, pressure vessel, floors, building walls, and storage pools. Results were found to be consistent with the experimental scenario of a pressure vessel filled with water and with no nuclear fuel: the reactor was under maintenance during the periods of data-taking.

Following the encouraging results with the JAPC reactor, the same team investigated the status of the nuclear fuel assemblies in the Unit-1 reactor of the Fukushima Daiichi plant.[29] Because of the very high environmental radiation, reported to be $\sim$0.5 mSv/h near the Unit-1 reactor building and mainly consisting of gammas from Cs^{137} and Cs^{154}, the detection system was shielded with 10 cm of iron. The observation of the reactor inner structure was performed from the three locations: West-North (WN), North-West (NW), and North (N) with respect to the position of the Unit-1 reactor building. Figure 5 shows the reconstructed absorption map of the Unit-1 reactor from the WN position after 90 days of data-taking. The region enclosed by the red dashed line is where the reactor pressure vessel is located and shows no sign of the presence of heavy materials. This is consistent with the hypothesis that most of the nuclear fuel got melted and dropped from the vessel.

D. Pagano & A. Lorenzon

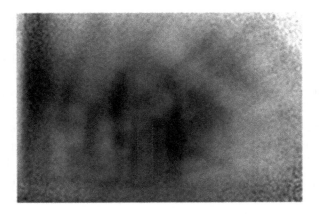

Figure 5. Muography of the Unit-1 reactor as reconstructed from the WN location after 90 days of data-taking. The red dashed line identifies the reactor pressure vessel.[29]

3.3. *Exploring the inner structure of buildings*

In addition to previous applications, where muography was used to provide a somehow coarse *view* of the inside of large and inaccessible structures, techniques of muon imaging have also been applied to the study of the structural components of buildings and industrial equipment, where a much fine resolution is required. An overview of such applications is given in Sections 3.3.1, 3.3.2, and 3.4.

3.3.1. *Search for a iron hoop inside the Florence Cathedral's Dome*

Since 2016, Guardincerri *et al.*[30,31] have been investigating the possibility to search for iron reinforcement elements inside the dome of Santa Maria del Fiore, by means of muon imaging. The *Cattedrale di Santa Maria del Fiore* is one of the most iconic buildings in Florence and is part of the UNESCO World Heritage Site. Despite numerous studies, the presence or not of an iron hoop within its dome has not been assessed yet, even though this knowledge is crucial to evaluate the safety of the structure. Previous investigations with metal detectors provided unsatisfactory results, because of the large thickness of the dome.

Figure 6. Left: Reconstructed image of the three iron bars inside the concrete mock-up wall, from a data-taking of 35 days. Right: Corresponding result from the Monte Carlo simulation.[30]

The authors, therefore, investigated the possibility of using the muon scattering tomography to assess the possible presence of any iron element. To prove the feasibility of this measurement, a concrete mock-up wall, incorporating three iron bars in it, was built. The concrete thickness (173 cm) was chosen to have the same equivalent thickness (in terms of effects on the muon flux) of the dome. Results are shown in Fig. 6: the image on the left is what has been experimentally reconstructed from a data-taking of 35 days, whereas the image on the right is what was expected from an MC simulation.

The authors proved the technique could effectively be used to find iron bars inside of a 173 cm thick concrete wall, making the method suitable for investigating the interior of the dome in Santa Maria del Fiore.

3.3.2. *Detection of rebars in concrete*

An accurate condition assessment of concrete structures, necessary for planning maintenance operations, is a crucial topic due to aging infrastructure. Such an assessment involves the search for rebars, cracks, voids, and other flaws. Non-destructive techniques, such as ground penetrating radars,[32] infrared thermographics,[33] and ultrasonics[34] have consistently been used for the task, even though they can suffer from limited penetration depth, resolution, and restrictions in detecting certain features.[35] Finally, even though high resolution, and high depth imaging can be achieved (at least in principle) by means of X-ray[36] and neutron radiography,[37] both techniques pose safety risks. In recent years, some authors have investigated the possibility of muon imaging to overcome some of the shortcomings in existing techniques.

Niederleithinger *et al.*[35] compared the performance of muography in detecting reinforcement bars in concrete with other state-of-the-art techniques, namely radar, ultrasound, and X-ray laminography. For the test, a reference concrete block, of size $1.2\,m \times 1.2\,m \times 0.2\,m$, was produced and filled with four different iron targets, which are typical for reinforced concrete structures, placed at three different depths. Figure 7 shows the reconstruction of the uppermost layer of 6 mm diameter reinforcement bars (corresponding to a depth of 5 cm) as obtained from muon tomography (exposure time = 1,203 h) (b), radar (c), ultrasound (d), and X-ray laminography (exposure time = 1,000 ms) (e). The image from muon tomography shows a better resolution than those from both ultrasound and radar (although the latter has a better clarity), but does not reach the quality of X-ray tomography. However, because of the safety risks associated with X-ray tomography, muon tomography could still represent a valid option in some scenarios, despite the much longer data-taking time compared to the other techniques.

Dobrowolska *et al.*[38] carried out extensive MC simulations to assess the muon scattering tomography capability to detect and locate rebars in concrete structures. Different geometries were simulated, reproducing typical reinforcement configurations in concrete. The authors could detect and locate rebars, with a diameter as small as $33.7 \pm 7.3\,mm$ and a length of 100 cm, within a $100\,cm \times 100\,cm \times 50\,cm$ concrete block. They also showed that 30 mm diameter rebars could be resolved as separate objects if separated by at least 40 mm (see Fig. 8(c)), and that it was possible to distinguish between single and double layer configurations of rebars with the proposed technique.

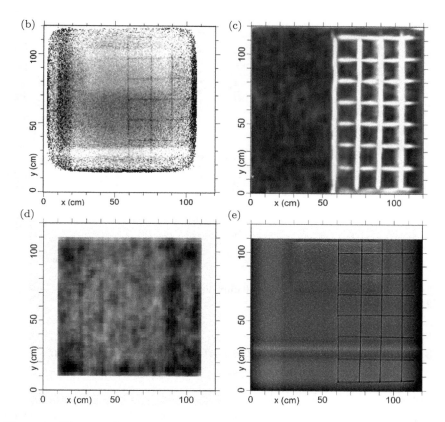

Figure 7. Horizontal cross-section, at depth of the uppermost layer of reinforcement bars (5 cm), as reconstructed by using: (b) muon tomography; (c) radar; (d) ultrasound; (e) X-ray laminography. Adapted from.[35]

Chaiwongkhot *et al.*[39,40] developed a compact muography detection system for measuring infrastructure degradation. It consists of two position sensitive detectors, of the area of 140 mm × 140 mm, made of plastic scintillating fibers, connected to multipixel photon counters. The detector was placed in the basement of a seven-story concrete building. The structural thickness profile of the building was reconstructed from the muon attenuation distribution. Results were found to be consistent with the structural drawings of the building and, according to the authors, represent the first steps toward an application of muography to the infrastructure degradation survey.

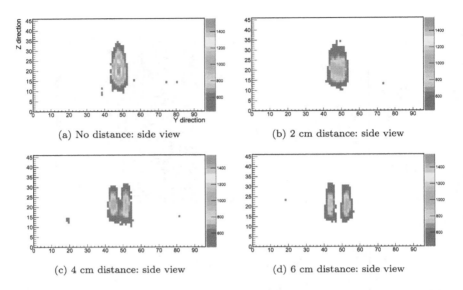

(a) No distance: side view

(b) 2 cm distance: side view

(c) 4 cm distance: side view

(d) 6 cm distance: side view

Figure 8. Muon imaging of two 30 mm diameter rebars separated by: 0 cm (a), 2 cm (b), 4 cm (c), and 6 cm (d).[38]

3.4. *Non-destructive testing of industrial equipment using muon radiography*

Martinez Ruiz del Arbol *et al.*[41] have proposed another possible application of muography, in the context of non-destructive testing of industrial equipment. The typically long-term operation of an industrial plant requires a continuous monitoring of its state of deterioration, in order to reduce the probability of a fault. In some cases, typically relevant to steel or petroleum industry, the corrosion of insulated equipment like pipes and cauldrons represents a serious problem, with high maintenance costs. By means of MC simulations, the authors investigated the possibility of using muography to assess the integrity of insulated equipment without removing the insulation layer, which is instead necessary with other techniques.[42] The idea was to measure the thickness of these components, also through the comparison with templates by means of a Kolmogorov–Smirnov test. Different pipes with outer radius of 20 cm, length 80 cm, and with inner radius varying from 5 to 19 cm were modeled. On the basis of the simulation results, produced with an equivalent exposure of ~6,900 s, the authors claimed the technique could identify changes in the inner radius as small as 0.2 cm.

4. Nuclear Controls

Ever since nuclear power has been used for electricity generation, nuclear waste has accumulated worldwide; one of the major challenges for the nuclear industry today is to safely store legacy, current, and future waste. Typical disposal strategies include encapsulating nuclear waste in different types of containers that act as a shield for the radiation emitted by the waste materials, typically neutron and gamma radiation. As a safeguard measure, the inspection of these containers is often required; however their shielding makes by its own nature the use of conventional imaging techniques — based on neutrons and/or gamma rays — really difficult.[43] In this context, muography has raised particular interest: in fact, due to their higher penetrating power, muons can be exploited as a non-destructive probe to identify and characterize the nuclear material inside these containers. Examples of the application of muon imaging techniques to nuclear waste containers will be given in Sections 4.1 and 4.2.

Other safeguard measures to counteract the diversion of nuclear material include the detection of the illicit transport of these materials, hidden in vehicles and cargo containers. As described in Section 4.3, MST has been proposed as a suitable technique for this purpose, since it is selective to high-Z materials. Another application related to transport controls, reported in Section 4.4, is the detection of radioactive sources that are occasionally present inside scrap metal containers, intended for steel recycling industries.

4.1. *Inspection of dry storage casks*

One of the most important waste management activities arising from the nuclear fuel chain is the handling of spent nuclear fuel: at the end of their life cycle inside nuclear reactors, spent fuel bars are highly radioactive and generate significant quantities of heat from

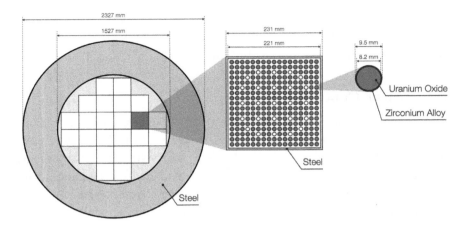

Figure 9. Schematic representation of cross-sections of a CASTOR®V/21, with details of a typical fuel assembly and a single fuel rod.

radioactive decay. Therefore, spent fuel has to be stored under water for several years in cooling pools that allow the decay heat to decrease while providing some shielding against radiation. Later, the common handling of spent fuel foresees the transfer of fuel bars inside dry storage casks (DSC), that are typically stored in dedicated spent fuel storage facilities.[44] A schematic representation of a model of DSC and a fuel assembly is displayed in Fig. 9.

Since spent fuel contains uranium and plutonium, the diversion of such nuclear material can have disastrous consequences. The Significant Quantity (SQ) is a measure to evaluate these consequences; one SQ corresponds to the sufficient quantity of nuclear material to produce an explosive device (approximately 8 kg of plutonium or 25 kg of highly enriched uranium).[45] One DSC can contain from 20 to 50 SQs and the total number of DSCs in the world is increasing continuously (there were more than 1,500 casks only in Europe at the beginning of 2019).[46] For nuclear safety purposes, DSCs must be monitored during their transfer to storage facilities and also during their long-term storage. In addition to current security measures (such as tamper-indicating seals applied to casks), for which failures cannot be excluded over the long storage time, it is desirable to have a prompt technique to check if the DSCs' contents are intact (re-verification).[46]

In this context, MST is emerging as a promising re-verification tool, since it is sensitive to the fuel assemblies inside casks and therefore it could detect the absence of SQs of nuclear material; furthermore, given the large amount of material in the DSC (both in its structure and in the stored spent fuel bars), the fraction of muons stopped by the cask is not negligible, so a muon radiography by absorption can also be performed.[4] As explained also in Chapter 3, both techniques require muon detectors to be placed around the cask to measure the incoming and, eventually, the outgoing muon trajectories.

Monte Carlo studies demonstrate the effectiveness of muography in recognizing the absence of one or more fuel assemblies from a cask. Several works, such as Paulson *et al.*[48] and Vanini *et al.*,[49] show that missing assemblies can be recognized with high statistical precision in a small amount of time (12–48 hours) using a muon detector that completely surrounds the DSC. Alternatively, a set of detectors with limited acceptance can be moved around the cask: in this way data collected from different viewing angles are combined to generate a tomographic image. Ancius *et al.*[47] reported the results from a GEANT4[50] simulation of a setup with two muon trackers of 60° angular acceptance positioned in the proximity of a specific DSC, called CASTOR®V/19[51] (as it can host up to 19 fuel assemblies): in this case, the simulated cask contains 16 fuel assemblies and 3 empty spots. 3D reconstructions of the simulated data (corresponding to ~ 48 hours of data-taking time) are performed with two algorithms, one based on the absorption of muons inside the object to be inspected and one on the measurement of the scattering angle; details of the two methods are given by Vanini *et al.*[49] and Schultz *et al.*,[52] respectively. Both techniques show that it is possible to locate missing fuel assemblies inside the cask, not only in the ideal case of a 360° detector coverage, but also with a data-taking configuration that foresees the rotation of the two detectors around the DSC. Examples of reconstructed images obtained in the latter case are shown in Fig. 10.

One of the main concerns in the application of muography techniques is the radioactivity escaping from DSC compared to

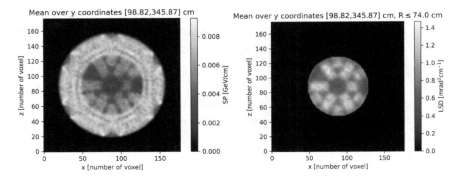

Figure 10. Reconstruction of a CASTOR® V/19 with three missing bars performed with the absorption technique (left) and with the scattering technique (right) on a simulated dataset corresponding to ∼48 hours of data-taking time, assuming a couple of detectors with 60° angular acceptance rotated around the cask in different positions. The external shield of the right figure is cut to avoid saturation.[47]

the low rate of cosmic muons. To assess the capability of a muon tracker to work in such a polluted environment, a small drift tube prototype (0.4 m × 2 m), built by National Institute of Nuclear Physics (INFN, Italy), has been tested in a radiation-free environment and in proximity of a CASTOR® container, loaded with spent fuel, in the interim storage facility of the EnKK nuclear power plant at Neckarwestheim (Germany).[53] The results confirmed the capability of the system to reconstruct muon tracks even in the presence of an important radioactivity, opening the path for further detector development and field tests.[47]

In 2016, a test performed at the Idaho National Laboratory (US) showed that the measurement of the scattering angles of cosmic ray muons through a partially loaded cask can be used to determine if some fuel assemblies are missing.[54] Two identical muon tracking detectors based on drift tubes were placed at the opposite sides of a Westinghouse MC-10 cask[55] (2.7 m large, 4.8 m high), filled with only 18 fuel bars out of 24; in order to intercept the muon flux at smaller zenith angles, one of the two detectors was raised 1.2 m above the other. Since the size of the detectors is small with respect to the cask (each detector covers 1.2 m × 1.2 m), both the upper and lower detectors were shifted horizontally in different positions to provide a good coverage of the container, as shown also on the left-hand side of Fig. 11. The measured scattering angles, collected during ~90 days of data-taking, are shown on the right-hand side of Fig. 11, as a function of the horizontal position across the cask and compared with GEANT4 simulations of a full and an empty cask. The result from the first column, that contains no fuel assembly, is consistent with the empty cask hypothesis, while the fourth column, with a single missing bar, displays an average scattering angle lower than what is expected for a full cask by 2.3 standard deviations.

In conclusion, simulations and data from the field tests so far show that muography is a promising technique for the re-verification of DSC.

4.2. *Imaging of intermediate level nuclear waste drums*

Low and intermediate level wastes, characterized by lower radioactivity with respect to spent nuclear fuel, are generated at all stages in the nuclear fuel cycle and also from the medical, industrial and research applications of radiation. The handling of these materials — for interim storage, for transport of final disposal — requires them to be enclosed in containers, to prevent any risk to the health and the environment. In particular, intermediate level wastes (ILWs) must be immobilized inside the container within a special cement structure (matrix), so that the waste and its container form a single monolithic body. This improves both the mechanical properties of the package and its ability to contain the radionuclides. Typical materials used for this purpose include, in addition to cements, also polymers, ceramics, and low melting point metal alloys.[56]

The ability to identify the content of these containers can mitigate the risks associated with long-term storage of nuclear waste, included containment breakdowns. In the particular case of legacy waste drums, individual vessels may date back a long time, and knowledge about the contents is not always preserved. Furthermore, both

D. Pagano & A. Lorenzon

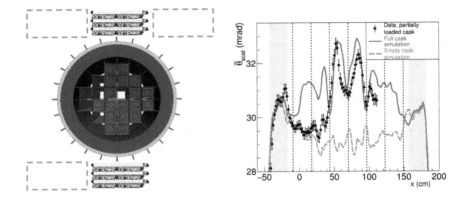

Figure 11. Left[a]: Representation of the MC-10 cask loading configuration and the positions of muon trackers during measurements. The wind was shaking the detector on the lower right position, so data taken in that configuration were not included. Right[a]: Average scattering angle as a function of the position across the cask. The gray areas and the vertical dashed lines indicate the boundaries of the shielding around the fuel and the boundaries of the fuel columns, respectively. The simulated scattering of a full (empty) cask is shown in blue (red).

chemical and physical processes inside the containers may have an affect on, and alter the composition of, the waste:[57] the oxidation of metals can lead to the formation of gas bubbles, which results in the expansion of the vessels and eventually cracks.

Since the seminal work of Borozdin *et al.*,[58] that indicated the potential of using the Coulomb scattering of muons for the identification of high atomic number objects hidden in a larger volume of material of low atomic number, MST has been proposed as a technology suitable for the imaging of ILW drums: in general it is required to identify high-Z materials (such as spent fuel fragments) inside these concrete-filled containers.

[a]Reprinted figure with permission from J. M. Durham, D. Poulson, J. Bacon, D. L. Chichester, E. Guardincerri, C.L. Morris, K. Plaud-Ramos, W. Schwendiman, J. D. Tolman, and P. Winston. *Phys. Rev. Applied* 9(4), 044013(8), (2018). URL: http://dx.doi.org/10.1103/PhysRevApplied.9.044013

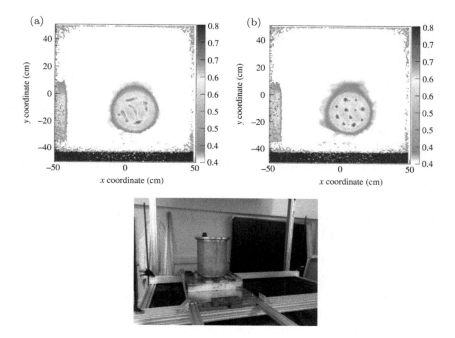

Figure 12. Top[b]: A 10 mm horizontal slice of the drum in a region with reinforcement steel bars (a) and 11 tungsten pennies, embedded in concrete. The high-density band at $y \leq -40$ cm is due to a faulty readout card, while the object on the left is part of the support structure. Images were reconstructed using data collected over a period of one month. Bottom[b]: Picture of the test container (300 mm large and 550 mm high) inside the muon imaging system.

In 2009, the University of Glasgow and the UK National Nuclear Laboratory started initial feasibility studies to assess the capability of MST to locate high-Z materials inside ILW stainless-steel concrete-filled drums. Simulations[59,60] have been performed to design a dedicated detector (a tracker with four detection modules, each containing orthogonal layers of plastic scintillating fibres), and

[b]Republished with permission from The Royal Society (U.K.), from "First-of-a-kind muography for nuclear waste characterization", D. Mahon, A. Clarkson, S. Gardner, D. Ireland, R. Jebali, R. Kaiser, M. Ryan, C. Shearer, and G. Yang, *Phil. Trans. R. Soc. A*, **377**(2137), 20180048, (2018); permissions conveyed through Copyright Clearance Center, Inc.

evaluate the first results obtained with high-Z samples. In 2015, the results of a small-scale vessel prototype were presented[61]: the clear discrimination between the steel casing, the concrete matrix, and the sample materials along with the good spatial resolution achieved confirmed that the MST technique is applicable to the inspection of sealed nuclear waste containers. The muon imaging system was further developed and commercialized by Lynkeos Technology.[62] In Fig. 12, the results of one of the validation tests performed with a 20 mm-thick stainless-steel drum filled with surrogate waste material are shown.[63]

4.3. Vehicle and cargo scanning for nuclear contraband

The detection of special nuclear materials (SNM) such as enriched uranium or plutonium that can be transported inside cargo containers has raised increasing attention in recent years, as the traffic of containers in transit across the world is continuously growing: monitoring the movement of these materials at key checkpoints such as country borders contributes to reducing the risk of nuclear attacks.

The traditional imaging techniques such as X-ray or γ-ray radiography are limited in the detection of shielded, hidden SNM and cannot be used for occupied vehicles due to radiation hazard to passengers. Inspection techniques based on multiple scattering of cosmic ray muons have been proposed as a new tool to detect high-Z materials in containers and trucks not only for their incomparable penetration and discrimination capabilities, but also because any radiation related risk would be removed. In Fig. 13, an example of detection of shielded radioactive material hidden in a truck obtained with a simulated muon portal is reported.

Several projects have focused on this application. For example, Gnanvo et al.[65] developed a small prototype with Gas Electron Multiplier (GEM) detectors and successfully tested it with medium and high-Z samples; Baesso et al.[66] studied the performances of a muon tracker based on high-resolution glass resistive plate chambers in order to develop an efficient and reliable scanner for shipping containers; Anghel et al.[67] reported the results obtained with a

Figure 13. Tomographic reconstruction of shielded SNM obtained with a simulated muon portal for truck inspection.[64]

scintillator-based MST system equipped with an integrated spectrometer for the measurement of the muon momentum, designed for the detection of nuclear materials inside a container (cargo, nuclear waste canister, etc.). More recently, the construction of a full-scale prototype for the inspection of cargo containers has been completed in Catania, Italy.[68]

This activity has also reached the commercial level: Decision Sciences International Corporation®[69] has developed a Multi-Mode Passive Detection System® (MMPDS),[70] a drive-thru portal for cargo containers and vehicles made up of aluminum drift tube arrays positioned above and below the volume of interest. The technology relies on the measurement of multiple Coulomb scattering and attenuation interactions of cosmic ray muons and electrons, that help to identify low-Z materials. The MMPDS also has gamma detection capability, so the presence of gamma-producing materials can be discovered. Typical scan times are around 1–2 minutes for clearing of benign cargo, while for suspicious configurations the scan time can be extended up to 10 minutes to gather more information. A $24' \times 72'$ muon portal from Decision Sciences is in operation in Freeport, Bahamas, since 2012.

4.4. *Search for radioactive sources in scrap metal*

Benettoni *et al.*[71] investigated the use of muography for the identification of radioactive sources in scrap metal, intended for the steel
industry. Indeed, a major concern exists that a radioactive source
(called "orphan" source) could be hidden inside the cargo with the
metal to be melted and, for this reasons, modern plants host radiation
scanning portals at the entrance. However, if the orphan source
is appropriately shielded, these devices fail in detecting it and its
consequent melting can result in severe environmental contamination
and economic loss.

To estimate the potential of muography in identifying orphan
sources in scrap metal, the authors put a $\sim$12.5 l lead block inside a
$1\,\mathrm{m}^3$ iron box filled with scrap metal, as shown on the left-hand side
of Fig. 14. The box, together with four reference blocks of iron and
lead, was placed inside a detection system[72] consisting of two 300 $\times$
$250\,\mathrm{cm}^2$ CMS barrel muon chambers.[73]

The tomographic reconstruction from a data taking of $\sim$20 min
is shown on the right-hand side of Fig. 14. A technique aiming at
monitoring vehicles entering an industrial plant is required to provide
a reliable response within a period of time of very few minutes, not

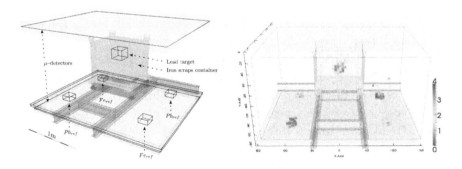

Figure 14. Left: Sketch of the experimental setup: a 25 $\times$ 25 $\times$ 20 cm^3 iron box,
filled with scrap metal and a lead target, was placed between 300 $\times$ 250 cm^2 drift
chambers. Four additional blocks (two of lead and two of iron) were placed near
the corners of the lower detectors as a reference. Right: Reconstructed image
based on experimental data (acquisition time $\sim$20 min). The geometry of the
considered setup was superimposed for clarity.[71]

to delay the industrial activities. Such a requirement is usually very challenging for muon imaging techniques, given the limited CR muon rate. However, the author showed that a full scale portal for truck inspection could make muography meet the requirement on duration of data taking.

5. Muon Metrology for Civil and Industrial Applications

In addition to the muon imaging applications already presented, it is also worth mentioning that some authors have investigated the possibility of using CR muons for the metrology of civil and industrial structures.[4]

Track-based alignment of detection systems, by means of CR muons, has a long history in nuclear and particle physics. Because of assembly tolerances and movements/deformations due to weight, temperature, and magnetic field, the geometry of a detector could substantially deviate from the original design. As a result, a dramatic degradation of tracking performance can occur. Techniques based on the reconstruction of charged particles, to infer the amount of misalignment in a detection system, are the most widespread, as they are sensitive to detector misalignment at levels smaller than the intrinsic detector resolutions.[74] Tracks from different sources have been used in track-based alignment, including CR muons.

In 2007, Bodini *et al.*[75] investigated the possibility of using a somehow analogous technique in a completely different scenario: the monitoring of the structural alignment of a mechanical press. The idea was to assess the potential misalignment between different parts of a large mechanical press ($5 \times 1.5 \times 1.2$ m^3), for which the presence of interposing material could prevent the use of conventional techniques. In their work, based on MC simulations, the authors considered a tracking system, consisting of three 20 cm $\times$ 20 cm position sensitive layers, integral to different parts of a mechanical press, as shown on the left-hand side of Fig. 15. The distance between the lower and upper layers was 3.3 m and the interposed iron accounted for a total thickness of 28.0 cm. The idea was to assess possible structural deformations of the press from the changes in the

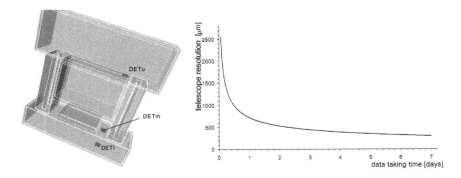

Figure 15. Left: Simulated configuration of the industrial press with the three position sensitive layers (DETu, DETm, and DETl). Right: Resolution on the relative shift between DETu and DETl as a function of the duration of data taking.[75]

relative misalignment of the sensitive layers. Results showed that a relative horizontal shift of ∼1 mm (∼0.5 mm) can be detected with a data taking of ∼12 h (∼60 h), as shown in Fig. 15, on the right.

In 2014, Zenoni *et al.*[76] proposed, for the first time, the use of muon metrology for the stability monitoring of historical buildings, where constraints related to their cultural and artistic value could prevent the use of standard monitoring systems. In a more recent work,[77] the authors developed further the idea and presented a feasibility study of the technique for the stability monitoring of the wooden dome of *Palazzo della Loggia* in Brescia, Italy. Like the case of the industrial press, the idea was to assess the structural modifications of the dome from the changes in the relative alignment over time of a set of tracking detectors, some of them integral to the structure of interest and the others to the surrounding environment. In their work, based on MC simulations, the authors considered two muon telescopes, consisting of three $400 \times 400 \, mm^2$ position sensitive elements each. Three different configurations were considered, with the detectors being vertically separated (Δz) by 350 cm, 880 cm, and 1,300 cm, respectively. Figure 16 shows the resolutions on the relative shift (top) and inclination (bottom) of the two telescopes: values of the order of 1 mm (shift) and 0.5 mrad (inclination) could be achieved

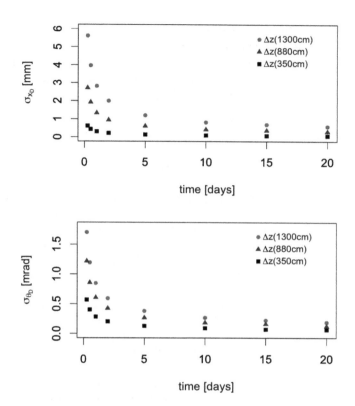

Figure 16. Resolution on the relative shift (top) and inclination (bottom) between the two telescopes as a function of time and for the three different configurations considered[77].

with few days of data takings even in the configuration with $\Delta z = 1,300$ cm. Such a performance is comparable to what is obtained from other monitoring systems, with the potentially appealing feature of applicability in presence of interposing structures and limited invasiveness. Moreover, as typical structural deformations evolve over a very long period of time, data takings of the order of few days do not represent a potential drawback of the technique, unlike other applications.

It is interesting to point out that, although previous applications of muon metrology are basically track-based alignment problems, an important difference exists:[4] while alignment procedures aim at providing a precise estimation of all alignment parameters, the muon metrology of structures, instead, focuses on how the alignment parameters change over time, regardless of their actual values. As a result, several sources of systematic uncertainties are less relevant for the muon metrology, as they could be canceled out.

References

1. E. George, Cosmic rays measure overburden of tunnel, *Commonwealth Engineer*, **455**, (1955).
2. S. Bouteille, D. Attié, P. Baron, D. Calvet, P. Magnier, I. Mandjavidze, S. Procureur, M. Riallot, and M. Winkler, A micromegas-based telescope for muon tomography: The WatTo experiment, *Nuclear Instruments and Methods in Physics Research Section A: Accelerators, Spectrometers, Detectors and Associated Equipment*, **834**, 223–228, (Oct., 2016). doi:10.1016/j.nima.2016.08.002. https://doi.org/10.1016/j.nima.2016.08.002
3. S. Procureur, Muon imaging: Principles, technologies and applications, *Nuclear Instruments and Methods in Physics Research Section A: Accelerators, Spectrometers, Detectors and Associated Equipment*, **878**, 169–179, (Jan., 2018). doi:10.1016/j.nima.2017.08.004. https://doi.org/10.1016/j.nima.2017.08.004
4. G. Bonomi, P. Checchia, M. D'Errico, D. Pagano, and G. Saracino, Applications of cosmic-ray muons, *Progress in Particle and Nuclear Physics*. **112**, 103768, (May, 2020). doi:10.1016/j.ppnp.2020.103768. https://doi.org/10.1016/j.ppnp.2020.103768
5. J. Lowe, and P. F. Zaccheo, Subsurface explorations and sampling. In *Foundation Engineering Handbook*, pp. 1–71. Springer US (1991). doi:10.1007/978-1-4615-3928-5_1. https://doi.org/10.1007/978-1-4615-3928-5_1
6. K. V. Ragnarsdóttir, Rare metals getting rarer, *Nature Geoscience*, **1**(11), 720–721, (Nov., 2008). doi:10.1038/ngeo302. https://doi.org/10.1038/ngeo302
7. T. Takahashi, ISRM suggested methods for land geophysics in rock engineering, *International Journal of Rock Mechanics and Mining Sciences*. **41**(6), 885–914, (Sept., 2004). doi:10.1016/j.ijrmms.2004.02.009. https://doi.org/10.1016/j.ijrmms.2004.02.009
8. Z.-X. Zhang, T. Enqvist, M. Holma, and P. Kuusiniemi, Muography and its potential applications to mining and rock engineering, *Rock Mechanics and Rock Engineering*, **53**(11), 4893–4907, (July, 2020). doi:10.1007/s00603-020-02199-9. https://doi.org/10.1007/s00603-020-02199-9

9. D. Bryman, J. Bueno, K. Davis, V. Kaminski, Z. Liu, D. Oldenburg, M. Pilkington, and R. Sawyer, Muon geotomography — bringing new physics to orebody imaging, *Building Exploration Capability for the 21st Century*, pp. 235–241 (2014).
10. D. Schouten, Muon geotomography: Selected case studies, *Philosophical Transactions of the Royal Society A*, **377**(2137), 20180061, (2019).
11. D. Bryman, J. Bueno, and J. Jansen, Blind test of muon geotomography for mineral exploration, *ASEG Extended Abstracts*, **2015**(1), 1–3, (2015).
12. P. D. Sloowere, B. Carlus, A. Chevalier, J.-. Ianigro, J. Marteau, D. Gilbert, and M. Rosas-Carbajal, How to detect disorders during tunnel digging with a muons telescope mounted on a TBM. In *24th European Meeting of Environmental and Engineering Geophysics*, European Association of Geoscientists & Engineers (2018). doi:10.3997/2214-4609.201802532. https://doi.org/10.3997/2214-4609.201802532
13. A. Chevalier, M. Rosas-Carbajal, B. Carlus, J.C. Ianigro, J. Marteau, D. Gibert, and P. De Sloovere, Monitoring muon flux variations besides and on a tunnel-boring machine. In *AGU Fall Meeting Abstracts*, vol. 2018, pp. NS23B–0704 (Dec., 2018).
14. A. Chevalier, M. Rosas-Carbajal, D. Gibert, A. Cohu, B. Carlus, J.C. Ianigro, F. Bouvier, and J. Marteau, Using mobile muography on board a Tunnel boring machine to detect man-made structures. In *AGU Fall Meeting Abstracts*, vol. 2019, pp. NS43B–0839 (Dec., 2019).
15. L.F. Thompson, J.P. Stowell, S.J. Fargher, C.A. Steer, K.L. Loughney, E.M. O'Sullivan, J.G. Gluyas, S.W. Blaney, and R.J. Pidcock, Muon tomography for railway tunnel imaging, *Physical Review Research*, **2**(2) (Apr., 2020). doi:10.1103/physrevresearch.2.023017. https://doi.org/10.1103/physrevresearch.2.023017
16. L. Oláh, G. Hamar, S. Miyamoto, H.K.M. Tanaka, and D. Varga, The first prototype of an MWPC-based borehole-detector and its application for muography of an underground pillar, *BUTSURI-TANSA(Geophysical Exploration)*, **71**(0), 161–168, (2018). doi:10.3124/segj.71.161. https://doi.org/10.3124/segj.71.161
17. D. Varga, G. Hamar, and G. Kiss, Asymmetric multi-wire proportional chamber with reduced requirements to mechanical precision, *Nuclear Instruments and Methods in Physics Research Section A: Accelerators, Spectrometers, Detectors and Associated Equipment*, **648**(1), 163–167, (Aug., 2011). doi:10.1016/j.nima.2011.05.049. https://doi.org/10.1016/j.nima.2011.05.049

18. A. Bonneville, R.T. Kouzes, J. Yamaoka, C. Rowe, E. Guardincerri, J.M. Durham, C.L. Morris, D.C. Poulson, K. Plaud-Ramos, D.J. Morley, J.D. Bacon, J. Bynes, J. Cercillieux, C. Ketter, K. Le, I. Mostafanezhad, G. Varner, J. Flygare, and A.T. Lintereur, A novel muon detector for borehole density tomography, *Nuclear Instruments and Methods in Physics Research Section A: Accelerators, Spectrometers, Detectors and Associated Equipment*, **851**, 108–117, (Apr., 2017). doi:10.1016/j.nima.2017.01.023. https://doi.org/10.1016/j.nima.2017.01.023

19. J. Perry, *Advanced Applications of Cosmic-ray Muon Radiography*. The University of New Mexico (2013).

20. K. Nagamine, H.K.M. Tanaka, S.N. Nakamura, K. Ishida, M. Hashimoto, A. Shinotake, M. Naito, and A. Hatanaka, Probing the inner structure of blast furnaces by cosmic-ray muon radiography, *Proceedings of the Japan Academy, Series B*, **81**(7), 257–260, (2005). doi:10.2183/pjab.81.257. https://doi.org/10.2183/pjab.81.257

21. E. Åström, G. Bonomi, I. Calliari, P. Calvini, P. Checchia, A. Donzella, E. Faraci, F. Forsberg, F. Gonella, X. Hu, J. Klinger, L.S. Ökvist, D. Pagano, A. Rigoni, E. Ramous, M. Urbani, S. Vanini, A. Zenoni, and G. Zumerle, Precision measurements of linear scattering density using muon tomography, *Journal of Instrumentation*, **11**(07), P07010–P07010, (July, 2016). doi:10.1088/1748-0221/11/07/p07010. https://doi.org/10.1088/1748-0221/11/07/p07010

22. X. Hu, L. S. Ökvist, E. Åström, F. Forsberg, P. Checchia, G. Bonomi, I. Calliari, P. Calvini, A. Donzella, E. Faraci, F. Gonella, J. Klinger, D. Pagano, A. Rigoni, P. Zanuttigh, P. Ronchese, M. Urbani, S. Vanini, A. Zenoni, and G. Zumerle, Exploring the capability of muon scattering tomography for imaging the components in the blast furnace, *ISIJ International*, **58**(1), 35–42, (2018). doi:10.2355/isijinternational.isijint-2017-384. https://doi.org/10.2355/isijinternational.isijint-2017-384

23. W. Gilboy, P. Jenneson, and N. Nayak, Industrial thickness gauging with cosmic-ray muons, *Radiation Physics and Chemistry*, **74**(6), 454–458, (Dec., 2005). doi:10.1016/j.radphyschem.2005.08.007. https://doi.org/10.1016/j.radphyschem.2005.08.007

24. J.A. Wheeler, Some consequences of the electromagnetic interaction between μ–mesons and nuclei, *Reviews of Modern Physics*, **21**(1), 133, (1949).

25. G. Brumfiel, and D. Cyranoski, *Quake Sparks Nuclear Crisis* (2011).

26. K. Borozdin, S. Greene, Z. Lukić, E. Milner, H. Miyadera, C. Morris, and J. Perry, Cosmic ray radiography of the damaged cores of the fukushima reactors, *Physical Review Letters*, **109**(15), 152501, (2012).

27. H. Miyadera, K.N. Borozdin, S.J. Greene, Z. Lukić, K. Masuda, E.C. Milner, C.L. Morris, and J.O. Perry, Imaging fukushima daiichi reactors with muons, *AIP Advances*, **3**(5), 052133, (May, 2013). doi:10. 1063/1.4808210. https://doi.org/10.1063/1.4808210

28. H. Fujii, K. Hara, S. Hashimoto, F. Ito, H. Kakuno, S. Kim, M. Kochiyama, K. Nagamine, A. Suzuki, Y. Takada, Y. Takahashi, F. Takasaki, and S. Yamashita, Performance of a remotely located muon radiography system to identify the inner structure of a nuclear plant, *Progress of Theoretical and Experimental Physics*, **2013**(7), (July, 2013). doi:10.1093/ptep/ptt046. https://doi.org/10.1093/ptep/ptt046

29. H. Fujii, K. Hara, K. Hayashi, H. Kakuno, H. Kodama, K. Nagamine, K. Sato, S.-H. Kim, A. Suzuki, T. Sumiyoshi, K. Takahashi, F. Takasaki, S. Tanaka, and S. Yamashita, Investigation of the unit-1 nuclear reactor of fukushima daiichi by cosmic muon radiography, *Progress of Theoretical and Experimental Physics*, **2020**(4) (Apr., 2020). doi:10.1093/ptep/ptaa027. https://doi.org/10.1093/ptep/ptaa027

30. E. Guardincerri, J.M. Durham, C. Morris, J.D. Bacon, T M. Daughton, S. Fellows, D.J. Morley, O.R. Johnson, K. Plaud-Ramos, D.C. Poulson, and Z. Wang, Imaging the inside of thick structures using cosmic rays, *AIP Advances*, **6**(1), 015213, (Jan., 2016). doi:10.1063/1.4940897. https://doi.org/10.1063/1.4940897

31. C. Blasi, L. Bonechi, R.D'Alessandro, E. Guardincerri, and V. Vaccaro, Detecting metal elements inside the dome of santa maria del fiore in florence using cosmic ray muons, *IOP Conference Series: Materials Science and Engineering*, **364**, 012072, (June, 2018). doi:10.1088/1757-899x/364/1/012072. https://doi.org/10.1088/1757-899x/364/1/012072

32. A. Tarussov, M. Vandry, and A.D.L. Haza, Condition assessment of concrete structures using a new analysis method: Ground-penetrating radar computer-assisted visual interpretation, *Construction and Building Materials*, **38**, 1246–1254, (Jan., 2013). doi:10.1016/j.conbuildmat. 2012.05.026. https://doi.org/10.1016/j.conbuildmat.2012.05.026

33. C. Maierhofer, R. Arndt, and M. Röllig, Influence of concrete properties on the detection of voids with impulse-thermography, *Infrared Physics & Technology*, **49**(3), 213–217, (Jan., 2007). doi:10.1016/j.infrared. 2006.06.007. https://doi.org/10.1016/j.infrared.2006.06.007

34. S. Laureti, M. Ricci, M. Mohamed, L. Senni, L. Davis, and D. Hutchins, Detection of rebars in concrete using advanced ultrasonic pulse compression techniques, *Ultrasonics*, **85**, 31–38, (Apr., 2018). doi:10.1016/j.ultras.2017.12.010. https://doi.org/10.1016/j.ultras.2017.12.010

35. E. Niederleithinger, S. Gardner, T. Kind, R. Kaiser, M. Grunwald, G. Yang, B. Redmer, A. Waske, F. Mielentz, U. Effner, C. Köpp, A. Clarkson, F. Thomson, M. Ryan, and D. Mahon, Muon tomography of the interior of a reinforced concrete block: First experimental proof of concept, *Journal of Nondestructive Evaluation*, **40**(3), (July, 2021). doi:10.1007/s10921-021-00797-3. https://doi.org/10.1007/s10921-021-00797-3

36. A. du Plessis and W.P. Boshoff, A review of x-ray computed tomography of concrete and asphalt construction materials, *Construction and Building Materials*, **199**, 637–651, (Feb., 2019). doi:10.1016/j.conbuildmat.2018.12.049. https://doi.org/10.1016/j.conbuildmat.2018.12.049

37. P. Zhang, F.H. Wittmann, P. Lura, H.S. Müller, S. Han, and T. Zhao, Application of neutron imaging to investigate fundamental aspects of durability of cement-based materials: A review, *Cement and Concrete Research*, **108**, 152–166, (June, 2018). doi:10.1016/j.cemconres.2018.03.003. https://doi.org/10.1016/j.cemconres.2018.03.003

38. M. Dobrowolska, J. Velthuis, A. Kopp, M. Perry, and P. Pearson, Towards an application of muon scattering tomography as a technique for detecting rebars in concrete, *Smart Materials and Structures*, **29**(5), 055015, (Mar., 2020). doi:10.1088/1361-665x/ab7a3f. https://doi.org/10.1088/1361-665x/ab7a3f

39. K. Chaiwongkhot, T. Kin, H. Ohno, R. Sasaki, Y. Nagata, K. Kondo, and Y. Watanabe, Development of a portable muography detector for infrastructure degradation investigation, *IEEE Transactions on Nuclear Science*, **65**(8), 2316–2324, (2018).

40. K. Chaiwongkhot, T. Kin, R. Sasaki, H. Sato, Y. Nagata, T. Komori, and Y. Watanabe. A feaibility study of 3d cosmic-ray muon tomography with a portable muography detector. In *Proceedings of the Second International Symposium on Radiation Detectors and Their Uses (ISRD2018)*, Journal of the Physical Society of Japan (Jan., 2019). doi:10.7566/jpscp.24.011010. https://doi.org/10.7566/jpscp.24.011010

41. P.M.R. del Arbol, P.G. Garcia, C.D. Gonzalez, and A. OrioAlonso, Non-destructive testing of industrial equipment using muon radiography, *Philosophical Transactions of the Royal Society A: Mathematical, Physical and Engineering Sciences*, **377**(2137), 20180054, (Dec., 2018). doi:10.1098/rsta.2018.0054. https://doi.org/10.1098/rsta.2018.0054

42. M. Twomey, Inspection techniques for detecting corrosion under insulation, *Materials evaluation*, **55**(2), (1997).

43. K.-P. Ziock, G. Caffrey, A. Lebrun, L. Forman, P. Vanier, and J. Wharton. Radiation imaging of dry-storage casks for nuclear fuel.

In *IEEE Nuclear Science Symposium Conference Record, 2005*, vol. 2, pp. 1163–1167 (2005).

44. M. Besnard, M. Buser, I. Fairlie, G. MacKerron, A. Macfarlane, E. Matyas, Y. Marignac, E. Sequens, J. Swahn, B. Wealer, *et al.* The world nuclear waste report 2019. focus europe. URL https://worldnuc learwastereport.org (2020).

45. IAEA. Iaea safeguards glossary 2001 edition. International Nuclear Verification Series 3, International Atomic Energy Agency, Vienna (2002).

46. A. Darius, K. Aymanns, P. Checchia, F. Gonella, A. Jussofie, F. Montecassiano, M. Murtezi, P. Schwalbach, K. Schoop, S. Vanini, and G. Zumerle. Modelling of safeguards verification of spent fuel dry storage casks using muon trackers. In *ESARDA 41st Annual Meeting Symposium on Safeguards and Nuclear Material Management*, pp. 142–148 (2019).

47. D. Ancius, P. Andreetto, K. Aymanns, M. Benettoni, G. Bonomi, P. Calvini, L. Castellani, P. Checchia, E. Conti, F. Gonella, A. Jussofie, A. Lorenzon, F. Montecassiano, M. Murtezi, K. Schoop, M.Turcato, and G. Zumerle, Muon tomography for dual purpose casks (mutomca) project. In *INMM/Esarda Joint Annual Meeting* (2021). https://resources.inmm.org/annual-meeting-proceedings/ muon-tomography-dual-purpose-casks-mutomca-project

48. D. Poulson, J. Durham, E. Guardincerri, C. Morris, J. Bacon, K. Plaud-Ramos, D. Morley, and A. Hecht, Cosmic ray muon computed tomography of spent nuclear fuel in dry storage casks, *Nuclear Instruments and Methods in Physics Research Section A: Accelerators, Spectrometers, Detectors and Associated Equipment*, **842**, 48–53, (2017). ISSN 0168-9002. doi:https://doi.org/10.1016/j.nima.2016.10.040. https://www. sciencedirect.com/science/article/pii/S0168900216310786

49. S. Vanini, P. Calvini, P. Checchia, A. Rigoni Garola, J. Klinger, G. Zumerle, G. Bonomi, A. Donzella, and A. Zenoni, Muography of different structures using muon scattering and absorption algorithms, *Philosophical Transactions of the Royal Society A*, **377**(2137), 20180051, (2019).

50. S. Agostinelli *et al.*, GEANT4–a simulation toolkit, *Nuclear Instruments and Methods in Physics Research A*, **506**, 250–303, (2003). doi:10.1016/S0168-9002(03)01368-8.

51. https://www.gns.de/language=en/21551/castor-v-19

52. L.J. Schultz, G.S. Blanpied, K.N. Borozdin, A.M. Fraser, N.W. Hengartner, A.V. Klimenko, C.L. Morris, C. Orum, and M.J. Sossong, Statistical reconstruction for cosmic ray muon tomography, *IEEE Transactions on Image Processing Research*, **16**(8), 1985–1993, (2007). doi:10.1109/TIP.2007.901239.

53. P. Checchia *et al.* Muography of spent fuel containers for safeguard purposes: A feasibility test in proximity of a castor container. In *IAEA Symposium on International Safeguards*, number CN267-048 (2018).
54. J.M. Durham, D. Poulson, J. Bacon, D.L. Chichester, E. Guardincerri, C.L. Morris, K. Plaud-Ramos, W. Schwendiman, J.D. Tolman, and P. Winston, Verification of spent nuclear fuel in sealed dry storage casks via measurements of cosmic-ray muon scattering, *Phys. Rev. Applied*, **9**, 044013, (2018). doi:10.1103/PhysRevApplied.9.044013. https://link.aps.org/doi/10.1103/PhysRevApplied.9.044013
55. M.A. McKinnon, J.M. Creer, C.L. Wheeler, J.E. Tanner, D.P. Batalo, D.A. Dziadosz, E.V. Moore, D.H. Schooner, M.F. Jensen, and J.H. Browder, The mc-10 pwr spent fuel storage cask: Testing and analysis. Technical Report NP-5268, Electric Power Research Institute Report (7, 1987). https://www.osti.gov/biblio/6179559
56. E. Goldfinch, Containers for packaging of solid low and intermediate level radioactive wastes — iaea technical reports series no. 355, (1993).
57. C. Paraskevoulakos, K. Hallam, and T. Scott, Grout durability within miniaturised intermediate level waste drums at early stages of interior volume expansion induced by encapsulated metallic corrosion, *Journal of Nuclear Materials*, **510**, 348–359, (2018). ISSN 0022-3115. doi: https://doi.org/10.1016/j.jnucmat.2018.08.028. https://www.scienced irect.com/science/article/pii/S0022311518303155
58. K.N. Borozdin, G.E. Hogan, C. Morris, W.C. Priedhorsky, A. Saunders, L.J. Schultz, and M.E. Teasdale, Radiographic imaging with cosmic-ray muons, *Nature*, **422**(6929), 277–277, (2003).
59. A. Clarkson, D. Hamilton, M. Hoek, D. Ireland, J. Johnstone, R. Kaiser, T. Keri, S. Lumsden, D. Mahon, B. McKinnon, M. Murray, S. Nutbeam-Tuffs, C. Shearer, C. Staines, G. Yang, and C. Zimmerman, The design and performance of a scintillating-fibre tracker for the cosmic-ray muon tomography of legacy nuclear waste containers, *Nuclear Instruments and Methods in Physics Research Section A: Accelerators, Spectrometers, Detectors and Associated Equipment*, **745**, 138–149, (2014). ISSN 0168-9002. doi:https://doi.org/10.1016/j.nima.2014.01.062. https://www.science direct.com/science/article/pii/S0168900214001314.
60. A. Clarkson, D. Hamilton, M. Hoek, D. Ireland, J. Johnstone, R. Kaiser, T. Keri, S. Lumsden, D. Mahon, B. McKinnon, M. Murray, S. Nutbeam-Tuffs, C. Shearer, C. Staines, G. Yang, and C. Zimmerman, Geant4 simulation of a scintillating-fibre tracker for the cosmic-ray muon tomography of legacy nuclear waste containers, *Nuclear Instruments and Methods in Physics Research Section A: Accelerators, Spectrometers, Detectors and*

Associated Equipment, **746**, 64–73, (2014). ISSN 0168-9002. doi: https://doi.org/10.1016/j.nima.2014.02.019. https://www.sciencedire ct.com/science/article/pii/S0168900214001752.

61. A. Clarkson, D. Hamilton, M. Hoek, D. Ireland, J. Johnstone, R. Kaiser, T. Keri, S. Lumsden, D. Mahon, B. McKinnon, *et al.*, Characterising encapsulated nuclear waste using cosmic-ray muon tomography, *Journal of Instrumentation*, **10**(03), P03020—P03020, (Mar, 2015). ISSN 1748-0221. doi:10.1088/1748-0221/10/03/p03020. http://dx.doi.org/10.1088/1748-0221/10/03/P03020

62. https://www.lynkeos.co.uk

63. D. Mahon, A. Clarkson, S. Gardner, D. Ireland, R. Jebali, R. Kaiser, M. Ryan, C. Shearer, and G. Yang, First-of-a-kind muography for nuclear waste characterization, *Philosophical Transactions of the Royal Society A*, **377**(2137), 20180048, (2019).

64. Muons scanner to detect radioactive sources hidden in scrap metal containers (MU-STEEL). https://op.europa.eu/s/sPBK Research Fund for Coal and Steel RFSR-CT-2010-00033.

65. K. Gnanvo, L.V. Grasso, M. Hohlmann, J.B. Locke, A. Quintero, and D. Mitra, Imaging of high-z material for nuclear contraband detection with a minimal prototype of a muon tomography station based on gem detectors, *Nuclear Instruments and Methods in Physics Research Section A: Accelerators, Spectrometers, Detectors and Associated Equipment*, **652**(1), 16—20, (Oct, 2011). ISSN 0168-9002. doi:10.1016/ j.nima.2011.01.163. http://dx.doi.org/10.1016/j.nima.2011.01.163.

66. P. Baesso, D. Cussans, C. Thomay, J. J. Velthuis, J. Burns, C. Steer, and S. Quillin, A high resolution resistive plate chamber tracking system developed for cosmic ray muon tomography, *Journal of Instrumentation*. **8**(08), P08006–P08006, (2013). doi:10.1088/1748-0221/8/08/p08006. https://doi.org/10.1088/1748-0221/8/08/p08006

67. V. Anghel, J. Armitage, F. Baig, K. Boniface, K. Boudjemline, J. Bueno, E. Charles, P.-L. Drouin, A. Erlandson, G. Gallant, R. Gazit, D. Godin, V. Golovko, C. Howard, R. Hydomako, C. Jewett, G. Jonkmans, Z. Liu, A. Robichaud, T. Stocki, M. Thompson, and D. Waller, A plastic scintillator-based muon tomography system with an integrated muon spectrometer, *Nuclear Instruments and Methods in Physics Research Section A: Accelerators, Spectrometers, Detectors and Associated Equipment*, **798**, 12–23, (2015). ISSN 0168-9002. doi:https://doi.org/10.1016/j.nima.2015.06.054. https://www. sciencedirect.com/science/article/pii/S0168900215008049

68. F. Riggi, V. Antonuccio, M. Bandieramonte, U. Becciani, G. Bonanno, D. Bonanno, D. Bongiovanni, P. Fallica, G. Gallo, S. Garozzo, A. Grillo, P.L. Rocca, E. Leonora, F. Longhitano, D.L. Presti,

D. Marano, N. Randazzo, O. Parasole, C. Petta, S. Riggi, G. Romeo, M. Romeo, G. Russo, G. Santagati, M. Timpanaro, and G. Valvo, The muon portal project: Commissioning of the full detector and first results, *Nuclear Instruments and Methods in Physics Research Section A: Accelerators, Spectrometers, Detectors and Associated Equipment*, **912**, 16–19, (2018). ISSN 0168-9002. doi:https://doi.org/10.1016/j.nima.2017.10.006. https://www.science direct.com/science/article/pii/S0168900217310434 New Developments In Photodetection 2017.

69. https://decisionsciences.com
70. G. Blanpied, S. Kumar, D. Dorroh, C. Morgan, I. Blanpied, M. Sossong, S. McKenney, and B. Nelson, Material discrimination using scattering and stopping of cosmic ray muons and electrons: Differentiating heavier from lighter metals as well as low-atomic weight materials, *Nuclear Instruments and Methods in Physics Research Section A: Accelerators, Spectrometers, Detectors and Associated Equipment*, **784**, 352–358, (2015). ISSN 0168-9002. doi: https://doi.org/10.1016/j.nima.2014.11.027. https://www.sciencedire ct.com/science/article/pii/S0168900214013151 Symposium on Radiation Measurements and Applications 2014 (SORMA XV).
71. M. Benettoni, G. Bettella, G. Bonomi, G. Calvagno, P. Calvini, P. Checchia, G. Cortelazzo, L. Cossutta, A. Donzella, M. Furlan, *et al.*, Noise reduction in muon tomography for detecting high density objects, *Journal of Instrumentation*, **8**(12), P12007, (2013).
72. S. Pesente, S. Vanini, M. Benettoni, G. Bonomi, P. Calvini, P. Checchia, E. Conti, F. Gonella, G. Nebbia, S. Squarcia, G. Viesti, A. Zenoni, and G. Zumerle, First results on material identification and imaging with a large-volume muon tomography prototype, *Nuclear Instruments and Methods in Physics Research Section A: Accelerators, Spectrometers, Detectors and Associated Equipment*, **604**(3), 738–746, (2009). doi:10.1016/j.nima.2009.03.017. https://doi.org/10.1016/j.nima.2009.03.017
73. C. Collaboration, S. Chatrchyan, G. Hmayakyan, V. Khachatryan, A. Sirunyan, W. Adam, T. Bauer, T. Bergauer, H. Bergauer, M. Dragicevic, *et al.*, The cms experiment at the cern lhc, *JInst*, **3**, S08004, (2008).
74. V. Blobel, C. Kleinwort, and F. Meier, Fast alignment of a complex tracking detector using advanced track models, *Computer Physics Communications*, **182**(9), 1760–1763 (2011). ISSN 0010-4655. doi:https://doi.org/10.1016/j.cpc.2011.03.017. https://www.sciencedi rect.com/science/article/pii/S0010465511001093. Computer Physics

Communications Special Edition for Conference on Computational Physics Trondheim, Norway, June 23–26, 2010.
75. I. Bodini, G. Bonomi, D. Cambiaghi, A. Magalini, and A. Zenoni, Cosmic ray detection based measurement systems: A preliminary study, *Measurement Science and Technology*, **18**(11), 3537–3546, (Oct, 2007). doi:10.1088/0957-0233/18/11/038. https://doi.org/10.1088/0957-0233/18/11/038
76. A. Zenoni, G. Bonomi, A. Donzella, M. Subieta, G. Baronio, I. Bodini, D. Cambiaghi, M. Lancini, D. Vetturi, O. Barnabà, F. Fallavollita, R. Nardò, C. Riccardi, M. Rossella, P. Vitulo, and G. Zumerle, Historical building stability monitoring by means of a cosmic ray tracking system (2014).
77. G. Bonomi, M. Caccia, A. Donzella, D. Pagano, V. Villa, and A. Zenoni, Cosmic ray tracking to monitor the stability of historical buildings: A feasibility study, *Measurement Science and Technology*, **30**(4), 045901, (Mar., 2019). doi:10.1088/1361-6501/ab00d7. https://doi.org/10.1088/1361-6501/ab00d7

Ingram Content Group UK Ltd.
Milton Keynes UK
UKHW022150220323
418973UK00002B/42

9 789811 264900